THE DEEP-SEA CANYONS IN THE GULF OF GUINEA NEAR FERNANDO PÓO

FORAMINIFERAL FAUNAS FROM DEEP-SEA SEDIMENTS IN THE GULF OF GUINEA

Deel 30, 1973

Verhandelingen van het Koninklijk Nederlands geologisch mijnbouwkundig Genootschap

THE DEEP-SEA CANYONS IN THE GULF OF GUINEA NEAR FERNANDO PÓO
– J.J.H.C. Houbolt

FORAMINIFERAL FAUNAS FROM DEEP-SEA SEDIMENTS IN THE GULF OF GUINEA
– J. Brouwer

ISBN 978-94-017-6901-3 ISBN 978-94-017-7014-9 (eBook)
DOI 10.1007/978-94-017-7014-9

CONTENTS

THE DEEP-SEA CANYONS IN THE GULF OF GUINEA NEAR FERNANDO PÓO

J.J.H.C. HOUBOLT[1)]

SUMMARY

Two canyons descending from the shelf west of the island of Fernando Póo were followed to the deep ocean. They appear to terminate 110 nautical miles (200 km) from the shelf break in a water depth of 2800 m. A little further downslope, a new channel is formed which continues to descend over a distance of about 400 nautical miles (700 km) to a water depth of 4700 m. The sediments in this channel were found to be covered with a thin layer of Holocene clays. Beneath this recent cover, the canyon bottom is predominantly sandy and the natural levees predominantly clayey. The canyon-bottom sands are clean, moderately well sorted and mostly medium sized. Their greatest surface distribution occurs near the end of the channel where it widens before petering out at the lower end of the continental rise. No sands were found in the cores from the adjoining abyssal plain.

The majority of the sands have not been brought into the area by the Niger river, but have been derived from a region east of the Niger drainage area.

1. INTRODUCTION

In 1967/68, the Koninklijke/Shell Exploratie en Produktie Laboratorium carried out a deep-sea sedimentological survey in the Gulf of Guinea. The survey was conducted from a coastal freighter, the 'Admiraal Nelson', which was equipped with simple coring gear (H o u b o l t 1971), light seismic profiler equipment (single airgun or gas exploder, single-section hydrophone cable, and direct recording on a variable-density recorder), and a 12 kc echo-sounder.

The main purpose of the expedition was to study sedimentary processes in an area of active deposition of terrigenous deep-sea sediments.

2. MORPHOLOGY

2.1. *Introduction*

A detailed isobath chart of the area under consideration is given in figure 1, which shows the positions of all relevant tracklines and the topographic names used. Tracings of some echo-sounder profiles crossing the canyons are given in figure 2. The gradients of the canyons are shown in figure 3.

The Kwa Ibo Canyon and the Calabar Canyon could be followed in the deep sea to the 3.75 sec (2800 m) depth contour, where they appeared to terminate (figs. 1 and 2). The investigation of the course of these canyons was nevertheless continued because it was known that the Lamont Geological Observatory had taken a 6 m long sand core at the bottom of a large canyon some 250 miles to the west-southwest (H e e z e n, personal communication). It was then found that at about 3.85 sec (2875 m) water depth, a new channel was formed which continued some 400 nautical miles (700 km) downslope. We named this latter channel the "Principe Channel".

2.2. *Kwa Ibo and Calabar Canyons*

These canyons begin on the outer continental shelf south of Calabar and extend approximately 110 nautical miles seaward before apparently terminating. A l l e n (1964) suggests that the Calabar Canyon may connect with valleys developed on the surface of the late Pleistocene-early Holocene transgressive sands (Older sands) present on the Shelf in this area. Evidence for such a connection seems even stronger for the Kwa Ibo Canyon (fig. 2C in Allen 1964).

On the slope, both canyons have a maximum relief of about 180 m. Sounding control, however, is particularly poor in this area.

Seawards of the Niger Slope, the Kwa Ibo Canyon becomes the more prominent of the two, with a maximum relief of 0.12 sec (90 m) and well developed natural levees and terraces. The canyon is characteristically V-shaped in profile, but seawards of the 3.5. sec (2600 m) isobath it widens rapidly, decreases in relief, and finally can hardly be distinguished from the surrounding sea floor. The Calabar Canyon similarly widens and decreases in relief with increasing water depth, but near the end of its known extent it divides into two narrower channels.

2.3. *Principe Channel*

The Principe Channel is the largest component of the Fernando Póo Canyon System and extends for more than 400 nautical miles (700 km; fig. 1). It is first observed north-

[1)] Koninklijke/Shell Exploratie en Produktie Laboratorium Rijswijk, The Netherlands.

west of Principe as a number of small gullies that increase in relief seawards and unite to form a single channel (figs. 2 and 3).

This increase in relief is caused by an increase in gradient of the channel axis but not, apparently, by a change in the regional gradient (fig. 3). This suggests that the deepening of the gullies is not the result of a regional change of gradient.

The main Principe Channel reaches its maximum dimensions between depths of 4.0 sec (3000 m) and 5.0 sec (3800 m). Thereafter it decreases in width and relief until a depth of about 6.0. sec (4500 m), where there is an abrupt increase in width and a decrease in channel depth. These changes are accompanied by a decrease in the gradient of the channel and its sides (fig. 3). There is some evidence that near the apparent termination of the channel, its floor is higher than the adjacent sea floor beyond the natural levees. This suggests that here the channel is a depositional, rather than an erosional, feature.

Natural levees are a prominent topographic feature along much of the length of the Principe Channel, but appear to be absent along the tributary channels in the upper end of this channel. Small tributaries were found below a depth of 5 sec (3800 m). They do not descend from the Niger Rise or from the continental rise off Gabon, but they are believed to be formed by small gullies in the natural levees which unite and subsequently join the main channel. The irregular topography of the natural levees supports the view that many such small gullies exist.

The tributary channels at the upper end of the Principe Channel might originate from underflows leaving Calabar and Kwa Ibo Canyons. Trackline control does not exclude the possibility that a third tributary canyon joins the main channel from the east through a gap in the Guinea Ridge.

Distributary or diverging branches of the Principe Channel were not observed.

3. SEISMIC PROFILES

3.1. *Regional*

Regional seismic lines show that the Niger Cone and the continental rise towards the Guinea Ridge are underlain by at least 2 sec of undisturbed sediments. Basement reflections were only observed near the lower end of the Principe Channel.

3.2. *Kwa Ibo and Calabar Canyons*

Seismic resolution is generally too low to determine the details of the internal structures in the canyon walls and natural levees. On one section, however, it could be seen that the Calabar Canyon is associated with a depositional feature at a water depth slightly over 2 sec (1500 m), whereas the Kwa Ibo Canyon seems to be a purely erosional feature at a similar depth (fig. 4).

3.3. *Principe Channel*

3.3.1. *Main channel* (4-6 sec water depth)

Seismic sections across the Principe Channel between the depth contours of 4.4. and 5.1. sec (3290-3824 m) exhibit a zone of poorly defined reflectors approximately 0.15 sec thick, which overlies a zone of more strongly defined reflectors (fig. 5). This latter zone of strong reflectors is truncated at the base of the Principe Channel. In a core from station 32, the canyon-bottom sand was found to overlie a strongly tuffaceous clay. We therefore believe that the strong reflectors mentioned above are tuffaceous deposits derived from the nearby volcanoes of the Guinea Ridge.

In many cases a step down of about 0.05 sec can be observed on the top of the strong reflectors on both sides of the channel. On one side they are above the level of the canyon floor and on the other side they are at this level. This suggests migration of the channel towards the high side. The direction of the thus determined channel migration is given in figure 1. Further to the southwest, these strong reflectors occur below the level of the channel bottom.

3.3.2. *Lower end of Principe Channel*

At the far end, the Principe Channel widens and the natural levees diminish in height. The seismic sections here reveal a distinct contrast between the canyon floor and the natural levees (fig. 6). The channel floor is strongly reflecting and underlain by a series of other strong reflectors. Cores from this area revealed sands. The natural levees are acoustically transparent, often to such an extent that the sea floor is not resolved on the seismic records and has to be taken from the echo-sounder records. Cores in this latter area revealed the presence of muds.

Because of this pronounced difference, it was possible to chart the surface extent of the sand body at the end of the Principe Channel, and it was found to be about 800 km^2 in area.

About 0.2. sec below the natural levees and the canyon floor, a more wide-spread series of strong reflectors occurs (fig. 6). This suggests that the canyon floor and the natural levees have built upwards from this level, implying that the thickness of the sand body formed at the lower end of the Principe Channel would be about 200 m.

Correlation of the strongly reflecting horizon below the end of the Principe Channel with those truncated by the channel higher up seems possible. This would imply that these strong reflectors below the lower end of the Principe Channel are also tuffaceous clays.

4. DATA FROM CORES

4.1. *Introduction*

Several surface cores were taken, with a push corer of our own design (H o u b o l t 1971), from the canyon floor, the natural levees and the surrounding rise deposits. The maximum core length obtained was 2.9. m, but in sands the core recovery was poor because of sand leakage in the core catcher.

4.2. *Lithology*

The lithology of the cores obtained during the expedition was described by J.B.M. Jonker. His findings for the canyons near Fernando Póo are summarised below.

4.2.1. *Principe Channel*

All cores from the natural levees and the bottom of the Principe Channel exhibited a thin cover of silty to slightly silty clays of Holocene age. On the bottom of the channel (stations 28, 32, 39, 41 and V 19-285), Pleistocene sands were found beneath this clay cover. The sands are fine to medium and sometimes coarse grained, loosely packed and moderately well sorted (fig. 7). There is no systematic change in grain size going down the channel; small amounts of particles over 2 mm in size regularly occur.

The thickness of these sands is not known from cores. The longest sand core obtained was 6.18 m at station V 19-285 (by Lamont Geological Observatory; pers. comm., Heezen 1967) and only in the core from station 32 was the bottom of the sand found. Since this core was taken either on the canyon bottom near the wall or at the foot of the wall, it is no indication of the thickness of the major part of the sand body. The underlying lithology was found to be a strongly tuffaceous clay.

The core at station 39 exhibits large sand pockets in between silty clay. We have interpreted this as a slumping phenomenon of the nearby canyon wall.

The cores taken from the natural levees (stations 29, 34, 37 and 40) exhibit mainly clays, the upper parts being Holocene in age. These clays are silty to slightly silty and appear to be similar to those overlying the canyon-bottom sands. In the cores from stations 29, 37 and 40, sand layers were encountered. They are fine grained and very thinly laminated. At station 29, a 50 cm thick sand bed was found which graded from medium grained and moderately sorted at the bottom, to fine grained and well sorted at the top.

4.2.2. *Kwa Ibo Canyon* (station 27)

One core was taken in the canyon floor at the lower end of the Kwa Ibo Canyon. In this core, a 78 cm sandy clay layer was found to overlie fine to medium sand, in which some small-scale cross bedding was observed.

4.2.3. *Calabar Canyon* (stations 14 and 15)

The cores from station 14 on the canyon floor and station 15 on the natural levee consisted of homogeneous, slightly silty clay. No sand layers were found on the canyon floor.

4.2.4. *Conclusion*

From the above we conclude that sands occur most abundantly on the canyon floor of the Principe Channel and less frequently in the natural levees. These sands are covered with a thin veneer of clay. Hence there seems to be no modern sand transport through the Principe Channel. This view is corroborated by the study of foraminiferal content (see Section 4.3.).

No sand was found in cores from the Guinea abyssal plain. Hence in this case, as in Lake Geneva (Houbolt and Jonker 1968), the sand has been deposited mainly at the lower end of the rise.

The Calabar Canyon seems to be clay filled. Some sand was found in the floor of the Kwa Ibo Canyon. These latter statements are each based on one core only.

4.3. *Foraminifera*

The foraminiferal faunas of the cores obtained during the survey have been studied by J. Brouwer (this volume) and the results of his study, in so far as they are of interest for the present paper can be summarised as follows:

(i) The blanket of clays overlying the canyon-floor sands, as well as the upper part of the natural levees, is of Holocene age.
(ii) The sands in both the canyon floor and the natural levees are of Pleistocene age.
(iii) These sands contain an admixture of normal deep-sea faunas and displaced shallow-water faunas.

From this we conclude that:

1. Sand transport through the canyon was active during the Pleistocene when the coastline was close to the present-day shelf break.
2. During the post-glacial rise of the sea level, the coastline retreated. Little or no sand was transported over the wide shelves which then came into existence. Therefore only a thin covering blanket of clay was deposited during the Holocene.
3. The admixed, benthonic, shallow-water Foraminifera demonstrate the importance of the shallow-marine environment for the process of canyon sedimentation. Fluviatile sands did not move directly into the canyon heads, but remained in the shallow sea near the river mouth long enough to become admixed with the local Foraminifera before moving to the deep-sea.
4. The presence of a large admixture of a fauna common to deep-water environments indicates that the canyon sands did not move in one flow from the canyon head to the end of the canyon, but did so step by step and remained long enough at rest in between to become admixed with deep-sea foraminiferal assemblages.

4.4. *Bryozoa*

The Bryozoa content of the canyon deposits has been studied by Lagaaij (1973). His results can be summarised as follows:

1. Numerous shallow-water, nondeltaic Bryozoa were found in the canyon sands.

2. This displaced Bryozoa fauna is identical to the late Pleistocene faunas found in cores from shallow submarine outcrops on the Niger Shelf.
3. Based on the present-day water depth of these outcrops, on the original water depth in which these Bryozoa were living, and on the estimates of the duration of the post-glacial rise in sea level, the canyon sands sampled would have been deposited some 11000 years BP, which is the end of the Pleistocene.

4.5. *Heavy minerals*

The heavy-mineral assemblages found in the sands of the Principe Channel are characterised by a relatively high cyanite and amphibole content. For comparison, some sand samples from the Niger Shelf near the head of the Kwa Ibo and Calabar Canyons were analysed and found to contain a distinctly different assemblage rich in zircon, while the sand in the Kwa Ibo Canyon (station Nelson 27) contains an intermediate assemblage.

According to NEDECO (1961), most sand samples from the Niger River drainage area contain heavy-mineral assemblages rich in amphibole, while high percentages of zircon are characteristic for the eastern part of the delta. However, the most eastern sand samples from the Imo River area contain assemblages rich in cyanite. Cyanite-rich assemblages were also found in the Cameroun Shelf sands south-east of the Niger delta (Berthois, Crosnier and Le Calvez, 1968). The distribution of the heavy-mineral provinces is shown in figure 8. Table 1 shows the heavy mineral assemblages recognised.

Because of the cyanite-rich heavy mineral assemblage, we may conclude that the majority of the sands found in the Principe Channel were not brought in by the Niger. They have probably been derived from a region east of the Niger drainage area, either via the eastern part of the delta or via an unknown channel running down the continental rise off Cameroun or Rio Muno and passing through a gap in the Guinea Ridge north of Principe. However, we lack trackline control to check this latter possibility.

5. CONCLUSIONS

1. The morphology of the canyons and channels descending towards the deep ocean from the slopes near Fernando Póo is complex.
2. The canyon bottoms and the natural levees are overlain by a thin layer of Holocene clays. Hence there has been no recent sand transport through them.
3. Clean, fine to medium, sometimes coarse, usually moderately well sorted sands occur at the bottom of the Principe Channel below the Holocene clay cover.
4. All sand transport took place during Pleistocene low-sea-level stages when the coastlines were in the neighbourhood of the present-day shelf breaks.
5. The sands contain shallow-marine, near-coastal Foraminifera and Bryozoa. This indicates that they must have dwelt for some time in a near-coastal environment before being transported down the canyons.
6. The sands also contain a normal, deep-water assemblage of Foraminifera. This indicates that the sands descended step by step and dwelt on the ocean bottom in between these steps long enough to become enriched with the fauna typical of the deep-sea environment.
7. Heavy-mineral evidence indicates that the bulk of the sands was not brought in by the Niger River, but has been derived from a region east of the Niger drainage area.
8. At the lower end of the Principe Channel, the sandy channel bottom widens and a sand body with a surface area of about 800 km² is formed.
9. Seismic profiles suggest that the sand body is possibly 200 m thick. Some shale layers, such as the Holocene present-day clay cover, might occur however.

	Amphibole		Zircon		Disthene	
	RANGE	AVERAGE	RANGE	AVERAGE	RANGE	AVERAGE
● Amphibole assemblage	20-54%	35%	3-25%	12%	0-6%	2%
+ Zircon assemblage	0-23%	10%	22-71%	40%	0-8%	4%
o Cyanite assemblage	0-12%	3%	0-21%	7%	18-77%	40%
△ Amphibole-Zircon assemblage	22-28%	26%	19-28%	24%	0-9%	4%
▫ Zircon with Cyanite	2-24%	11%	25-34%	28%	11-21%	14%
✚ Cyanite-Amphibole assemblage	17-25%	21%	10-16%	13%	13-25%	21%

10. No sands were found in the adjoining abyssal plain.
11. There are indications of a lateral migration along the upper part of the Principe Channel. This would mean that the channel-bottom sands might occur in a sheet wider than the channel bottom.
12. The occurrence of a large clean Pleistocene sand body suggests that older, similar sand bodies can be expected to occur deeper down in the continental-margin sedimentary basins.

ACKNOWLEDGEMENT

Grateful acknowledgement is made to the author's colleagues at the Koninklijke/Shell Exploratie and Produktie Laboratorium for their contributions to this study, both in the laboratory and at sea.

REFERENCES

Allen, J.R.L. (1964) – The Nigerian continental margin: bottom sediments, submarine morphology and geologic evolution. Marine Geol., 1964, vol. 1, p. 289-332.

Berthois, L. A. Crosnier, Y. Le Clavez (1968) – Contribution à l'étude sédimentologique du plateau continental dans la baie de Biafra. Cah. ORSTOM Sér. Océanogr., 1969, vol. VI, nos. 3-4.

Brouwer, J. (1973) – Formaniniferal faunas from deep-sea sediments in the Gulf of Guinea. This volume.

Houbolt, J.J.H.C. (1971) – Transferable deep-sea coring gear. Marine Geol., 1971, vol. 10, p. 121-131.

Houbolt, J.J.H.C., J.B.M. Jonker (1968) – Recent sediments in the eastern part of the Lake of Geneva (Lac Léman). Geol. en Mijnbouw, vol. 47, no. 2, p. 131-148.

Lagaaij, R. (1973) – Shallow-water Bryozoa from deep-sea sands of the Principe Channel, Gulf of Guinea. In: Living and Fossil Bryozoa. Recent advances in research, G.P. Larwood Ed. Academic Press London 1973.

NEDECO (1961) – The waters of the Niger Delta. SRO The Hague, The Netherlands.

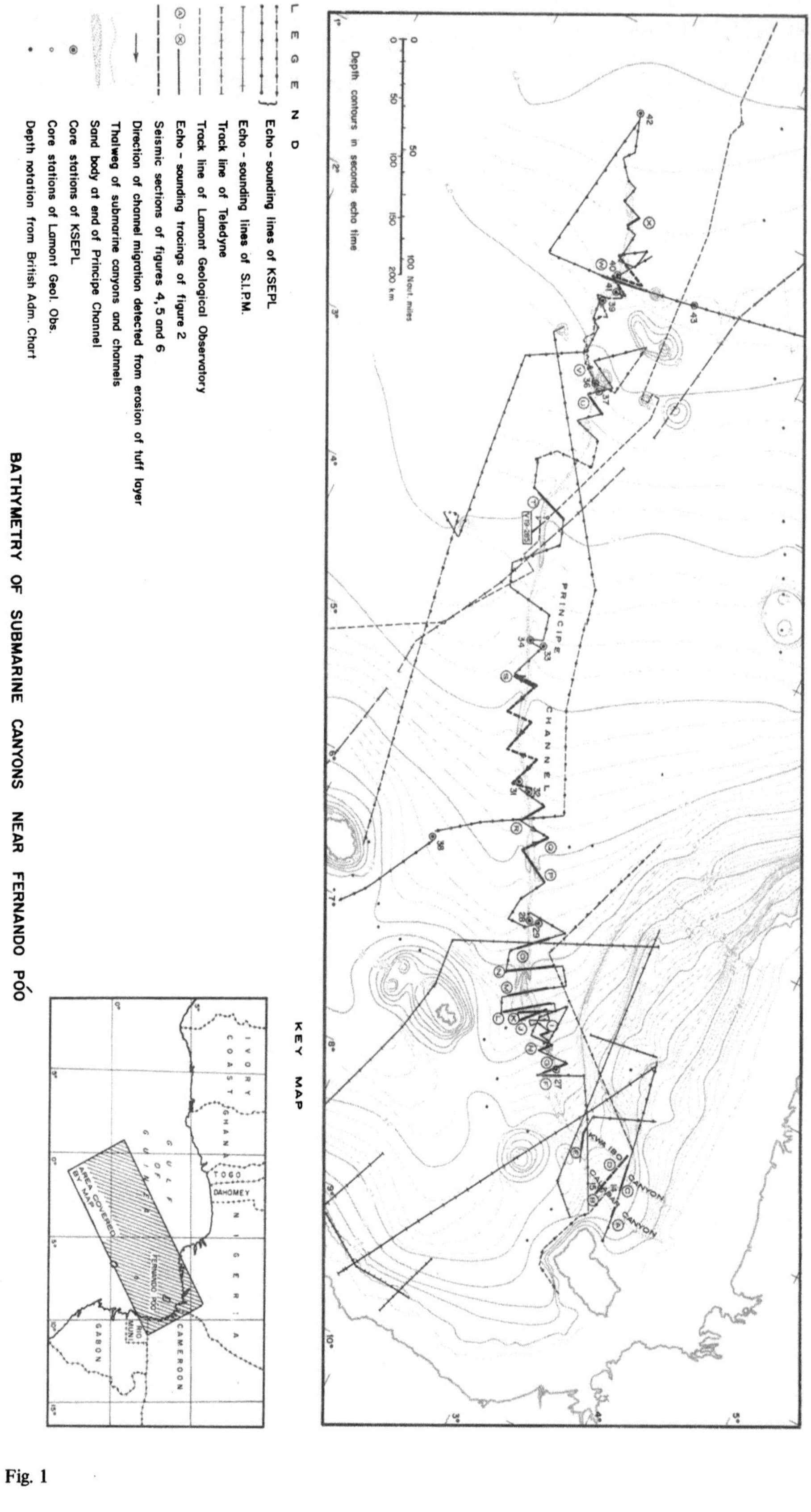
BATHYMETRY OF SUBMARINE CANYONS NEAR FERNANDO PÓO
LEGEND
Echo - sounding lines of KSEPL
Echo - sounding lines of S.I.P.M.
Track line of Teledyne
Track line of Lamont Geological Observatory
Echo - sounding tracings of figure 2
Seismic sections of figures 4, 5 and 6
Direction of channel migration detected from erosion of tuff layer
Thalweg of submarine canyons and channels
Sand body at end of Principe Channel
Core stations of KSEPL
Core stations of Lamont Geol. Obs.
Depth notation from British Adm. Chart
Depth contours in seconds echo time
100 Naut. miles
200 km
PRINCIPE CHANNEL
KEY MAP
IVORY COAST
GHANA
TOGO
DAHOMEY
NIGERIA
CAMEROON
GABON
RIO MUNI
GULF OF GUINEA
AREA COVERED BY MAP
CANYON

Fig. 1

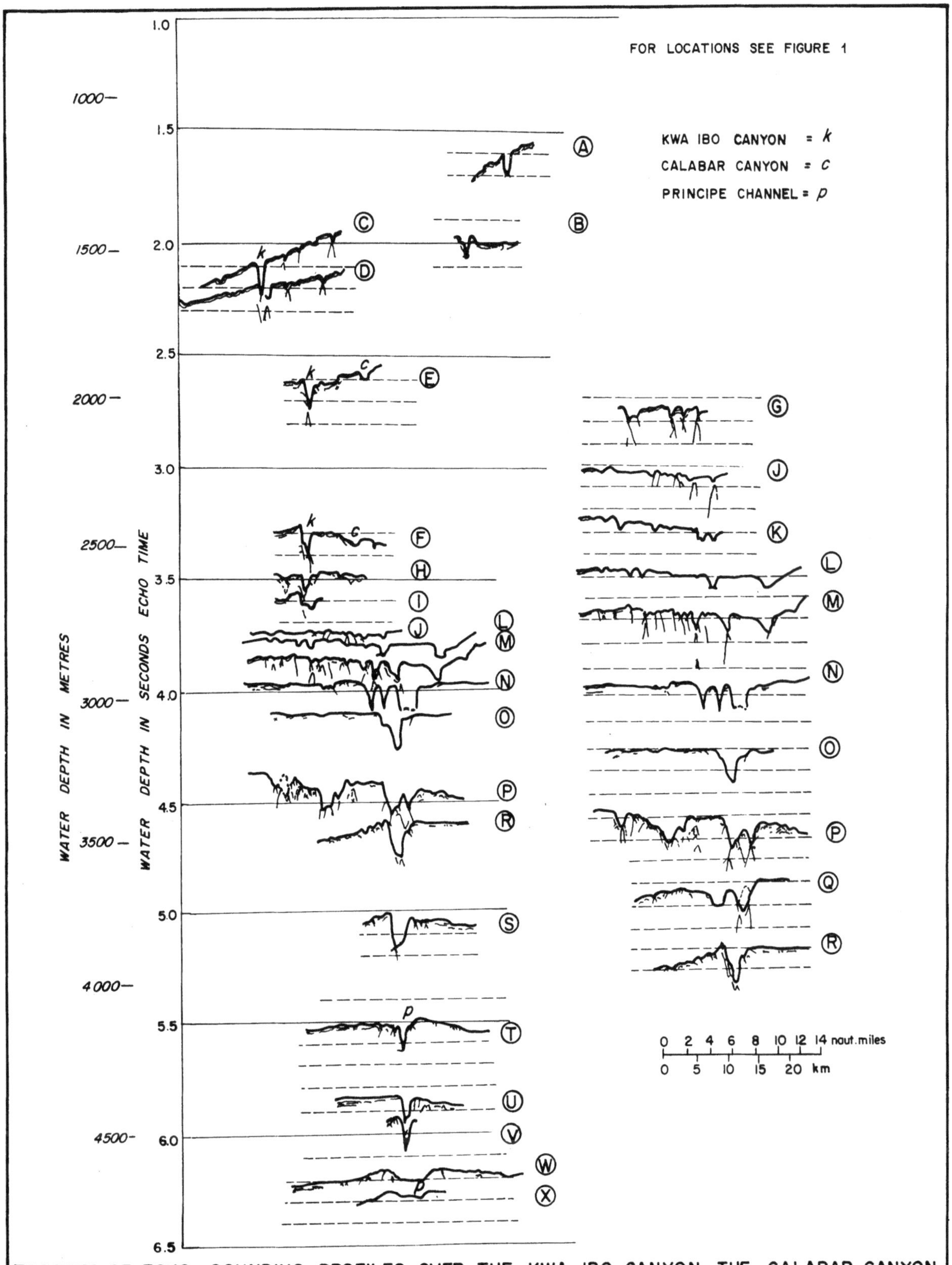

TRACINGS OF ECHO-SOUNDING PROFILES OVER THE KWA IBO CANYON, THE CALABAR CANYON AND THE PRINCIPE CHANNEL

Fig. 2

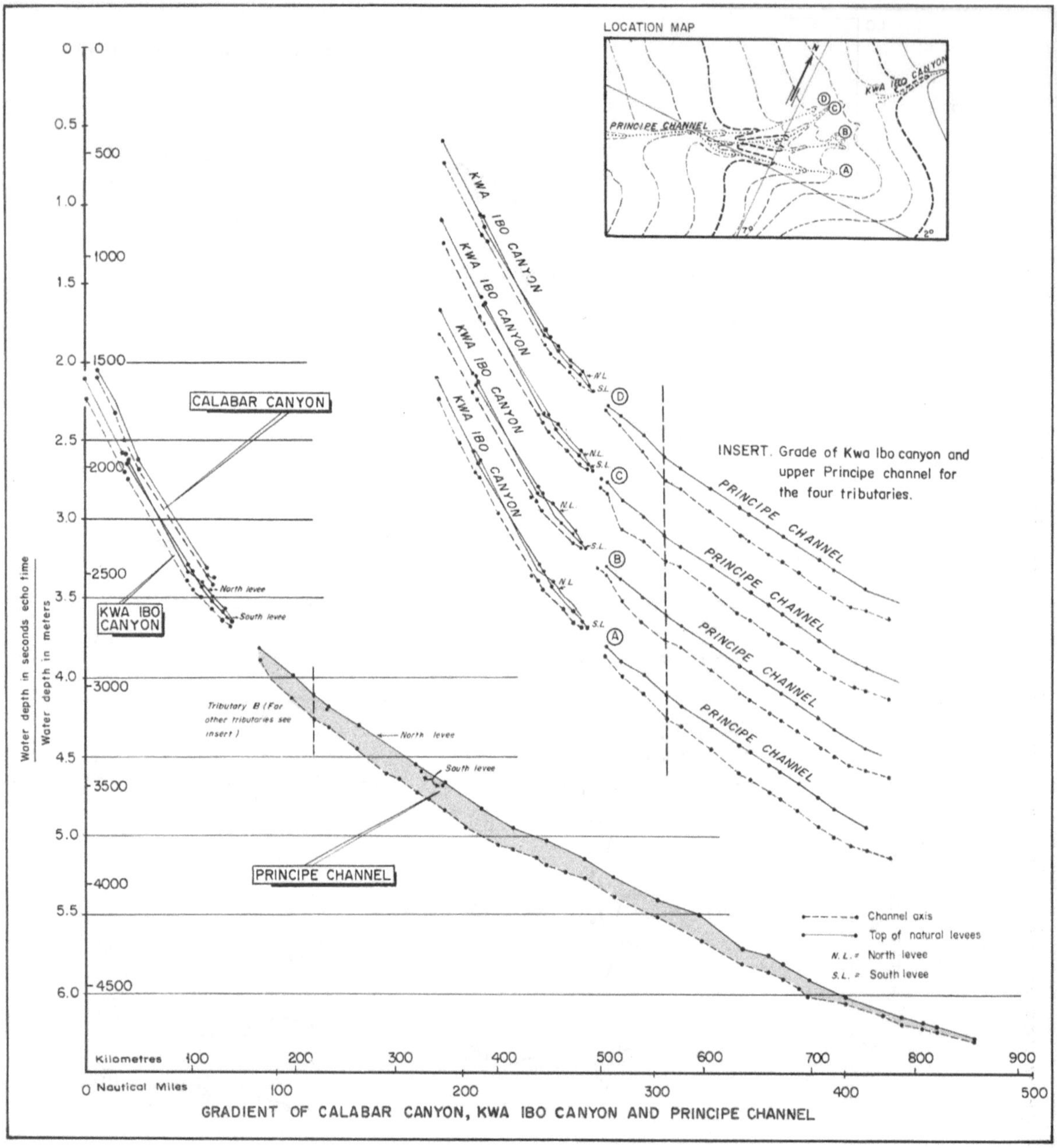
LOCATION MAP
PRINCIPE CHANNEL
KWA IBO CANYON
CALABAR CANYON
KWA IBO CANYON
PRINCIPE CHANNEL
INSERT. Grade of Kwa Ibo canyon and upper Principe channel for the four tributaries.
Tributary B (For other tributaries see insert)
North levee
South levee
Channel axis
Top of natural levees
N.L. = North levee
S.L. = South levee
Water depth in seconds echo time
Water depth in meters
Kilometres
Nautical Miles
GRADIENT OF CALABAR CANYON, KWA IBO CANYON AND PRINCIPE CHANNEL

Fig. 3

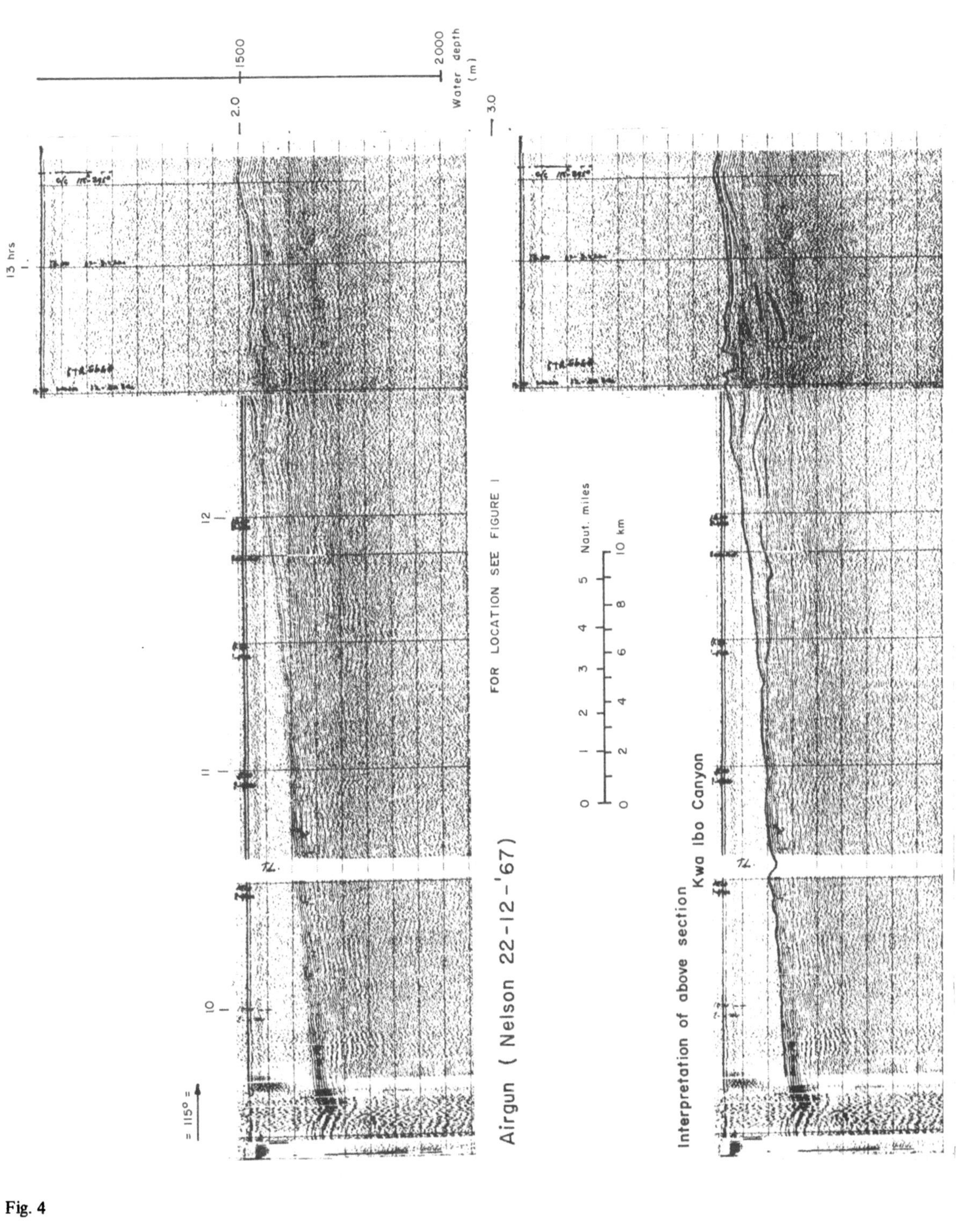
SEISMIC PROFILE OVER THE KWA IBO AND CALABAR CANYONS
Airgun (Nelson 22-12-'67)
Interpretation of above section
Kwa Ibo Canyon
FOR LOCATION SEE FIGURE 1
Naut. miles
0 1 2 3 4 5
0 2 4 6 8 10 km
13 hrs
12
11
10
= 115° =
2.0
3.0
1500
2000
Water depth
(m)

Fig. 4

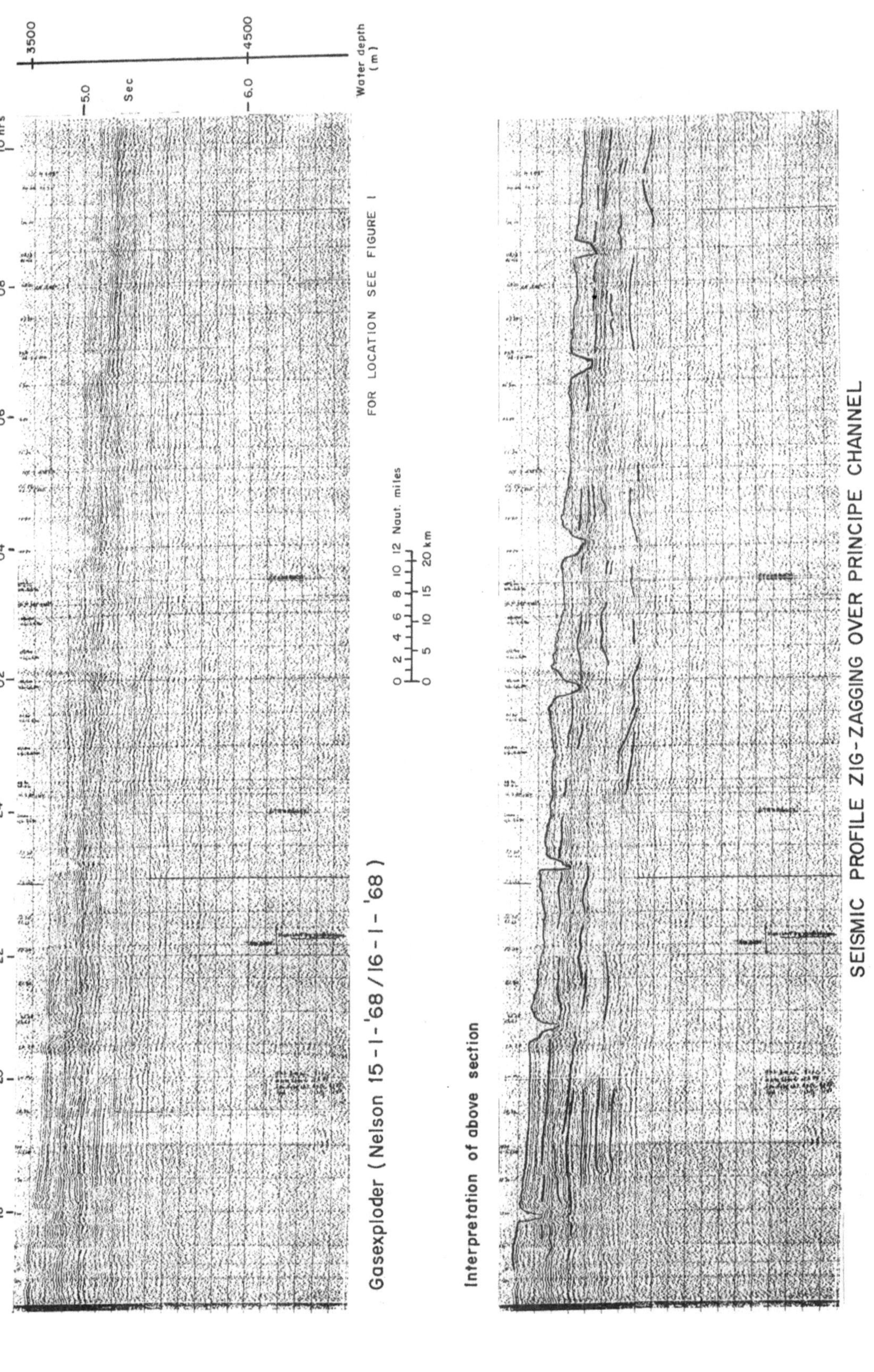

10 hrs
08
06
04
02
24
22
20
18
= 188° =
= 292° =
=188° =
15-1-'68 | 16-1-'68
= 293° =
=180° =
= 292° =
= 180° =
5.0
Sec
6.0
3500
4500
Water depth (m)
FOR LOCATION SEE FIGURE 1
0 2 4 6 8 10 12 Naut. miles
0 5 10 15 20 km
Gasexploder (Nelson 15-1-'68/16-1-'68)
Interpretation of above section
SEISMIC PROFILE ZIG-ZAGGING OVER PRINCIPE CHANNEL

Fig. 5

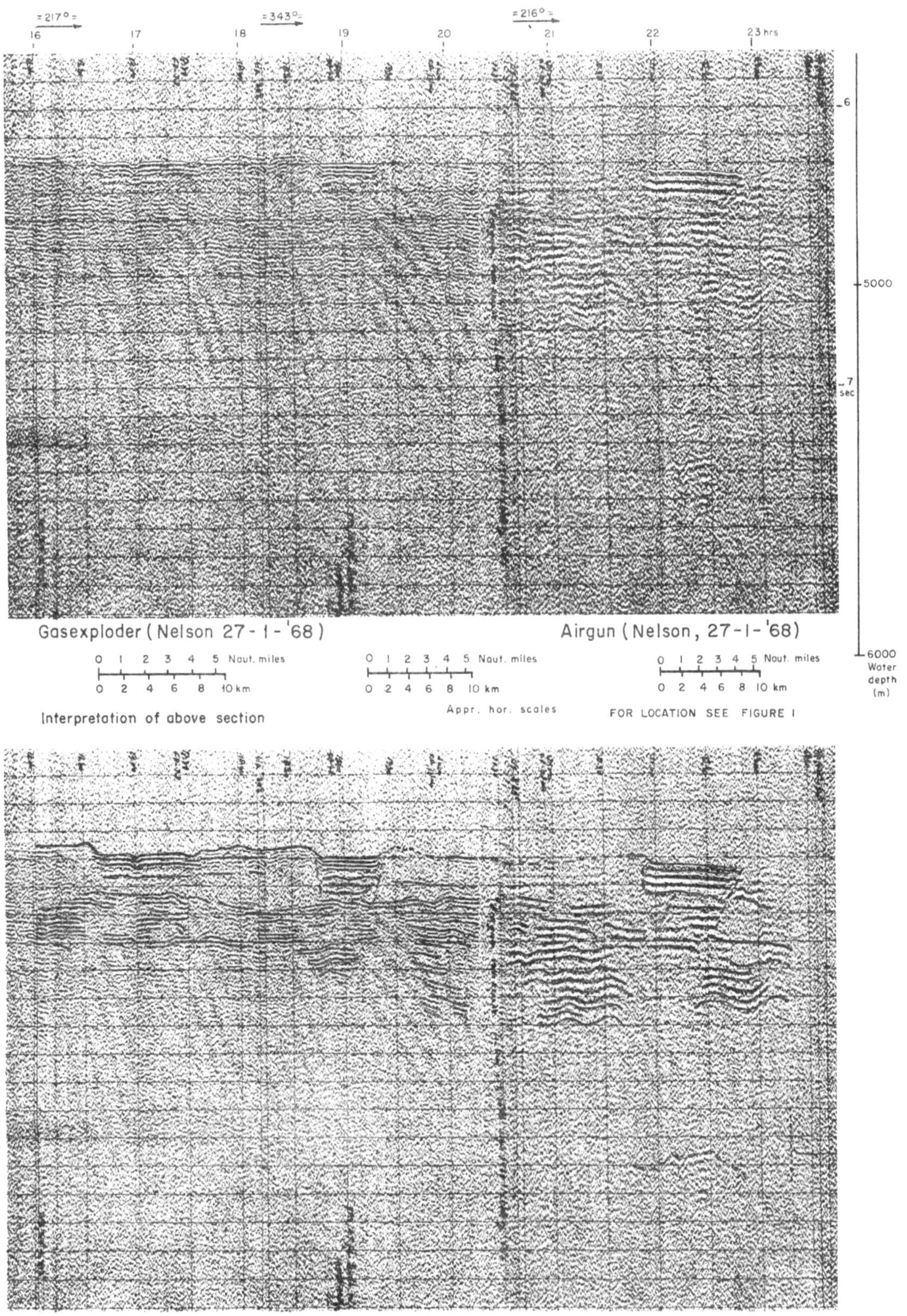

ZIG-ZAG SECTION OVER LOWER END OF PRINCIPE CHANNEL

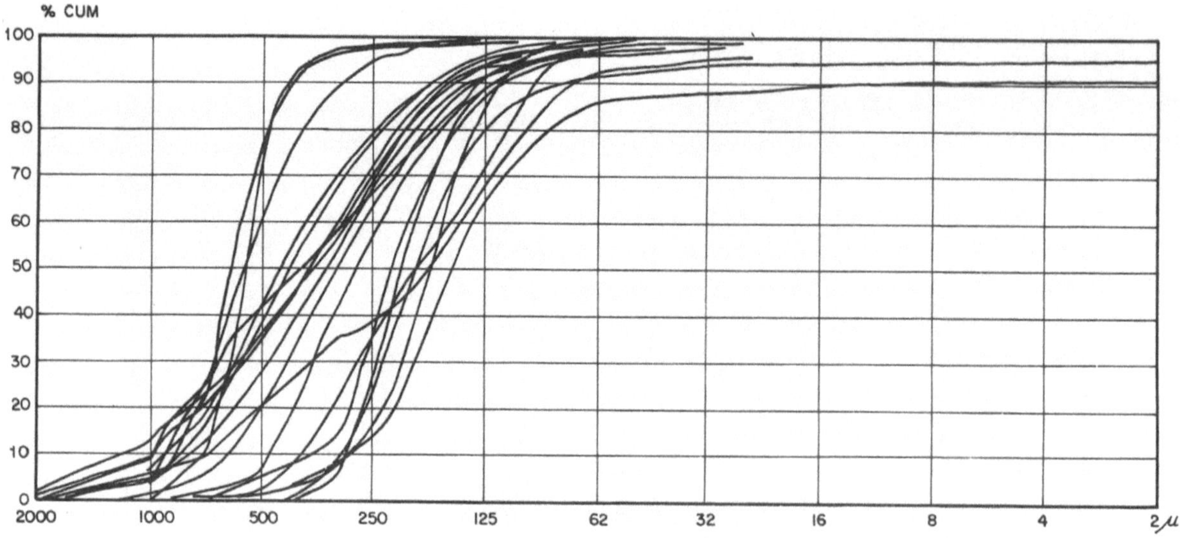

GRAIN - SIZE DISTRIBUTION OF SANDS FROM THE BOTTOM OF THE PRINCIPE CHANNEL.

Fig. 7

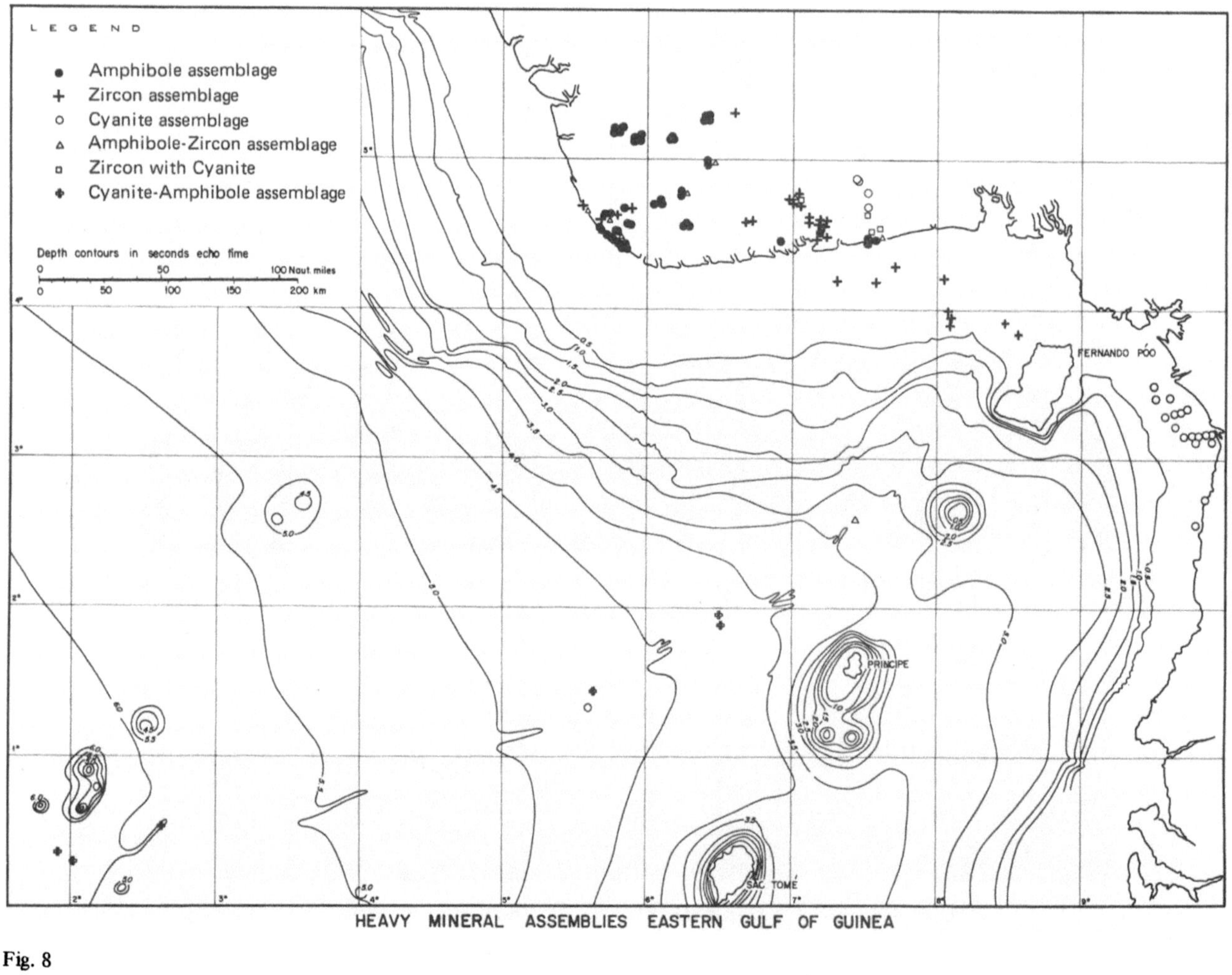

HEAVY MINERAL ASSEMBLIES EASTERN GULF OF GUINEA

Fig. 8

FORAMINIFERAL FAUNAS FROM DEEP-SEA SEDIMENTS IN THE GULF OF GUINEA

J. BROUWER[1)]

SUMMARY

Sediment cores taken during a deep-sea geological expedition in the Gulf of Guinea have been studied micropalaeontologically.

The fine-grained sediments below 1350 m sea depth contain a foraminiferal fauna, consisting of over 90% planktonic forms, accompanied by benthonic species characteristic of the Recent abyssal environments above the calcium carbonate compensation level.

Sandy layers bearing a high percentage of displaced shallow-water benthonic Foraminifera are often intercalated in these fine sediments in the submarine canyons and fan deposits at the canyon head.

Moreover, it became apparent that a thin blanket of Recent/Holocene sediments covers Pleistocene deposits of the same environment. Only one core penetrated into Pliocene strata.

I. INTRODUCTION

During 1967-1968 the Koninklijke/Shell Exploratie en Produktie Laboratorium undertook a deep-sea geological expedition in the Gulf of Guinea (Houbolt, 1973), in an area (see location map, text fig. 1) where, in principle, two types of sediment are found, i.e. pelagic sediments in the deeper part and terrigenous deposits in the shallower part (compare Sverdrup et al., 1946, p. 975).

This setting, combined with the presence of submarine canyons, offers an excellent opportunity for gathering more pertinent data on the composition of abyssal foraminiferal faunas, the degree of mixing caused by the transport of shallow neritic faunas down the canyons and the rate of sedimentation.

Such information could serve as a model for the interpretation of analogous fossil sediments.

In order to achieve this, 220 samples – chosen by Mr. J.B.M. Jonker to be representative of the lithological units present in each of 59 cores – were investigated both quantitatively and qualitatively for the presence of Foraminifera and other microfossil remains.

In the following paragraphs we discuss the method applied here, and give a description of the faunal characteristics and the conclusions reached. A bibliography and check list of foraminiferal species are also given at the end of the report.

[1)] Shell Internationale Petroleum Mij., The Hague, The Netherlands.

The author wishes to thank the directors of Shell Research N.V., The Hague, for permission to publish this article, and his colleagues, Dr. J.J.H.C. Houbolt and Dr. R. Lagaaij for their critical remarks and useful suggestions.

II. METHOD OF INVESTIGATION

Between 10 and 160 g of dry sediment from each sample were washed over a set of sieves, resulting in three fractions of wash-residue, i.e. 70-150 μ (fine), 150-420 μ (medium) and $>420>$ (coarse). These fine, medium and coarse fractions were then dried and weighed.

Next, the sieve fractions were investigated quantitatively for the presence of Foraminifera, Radiolaria, Echinoidea, Mollusca, Ostracoda, Bryozoa, and any remains of Pisces in the form of teeth and otoliths. In order to do this within a reasonable time, a sample splitter was used to obtain splits of residue in which between 100 and 200 specimens of both benthonic and planktonic Foraminifera could be expected.

The resulting specimen counts were then multiplied by the split factors for the purpose of calculating the number of foraminiferal specimens per gramme of sediment in each sample.

The percentage occurrence of each foraminiferal species, as found in the coarse and medium fractions combined, is presented on distribution charts (tables 1-6). These charts also indicate the core number, the sample interval, the approximate sea depth, the dry weight of the sample, and the weight-percentage of the three residues combined versus the dry weight of the sample.

In some samples which yielded insufficient specimens in the coarse and medium fractions, the percentage counts are based on all three fractions.

Also indicated on these tables are the total percentages of agglutinated and calcareous benthonic Foraminifera considered to be autochthonous, the total percentage of displaced Foraminifera, and the quantities per gramme of sediment of the non-foraminiferal faunal constituents.

On the distribution tables the percentages are rounded off to whole figures, half a percent or a higher fraction being one percent. Those occurrences of less than half a percent are indicated by +. For both planktonic and benthonic species

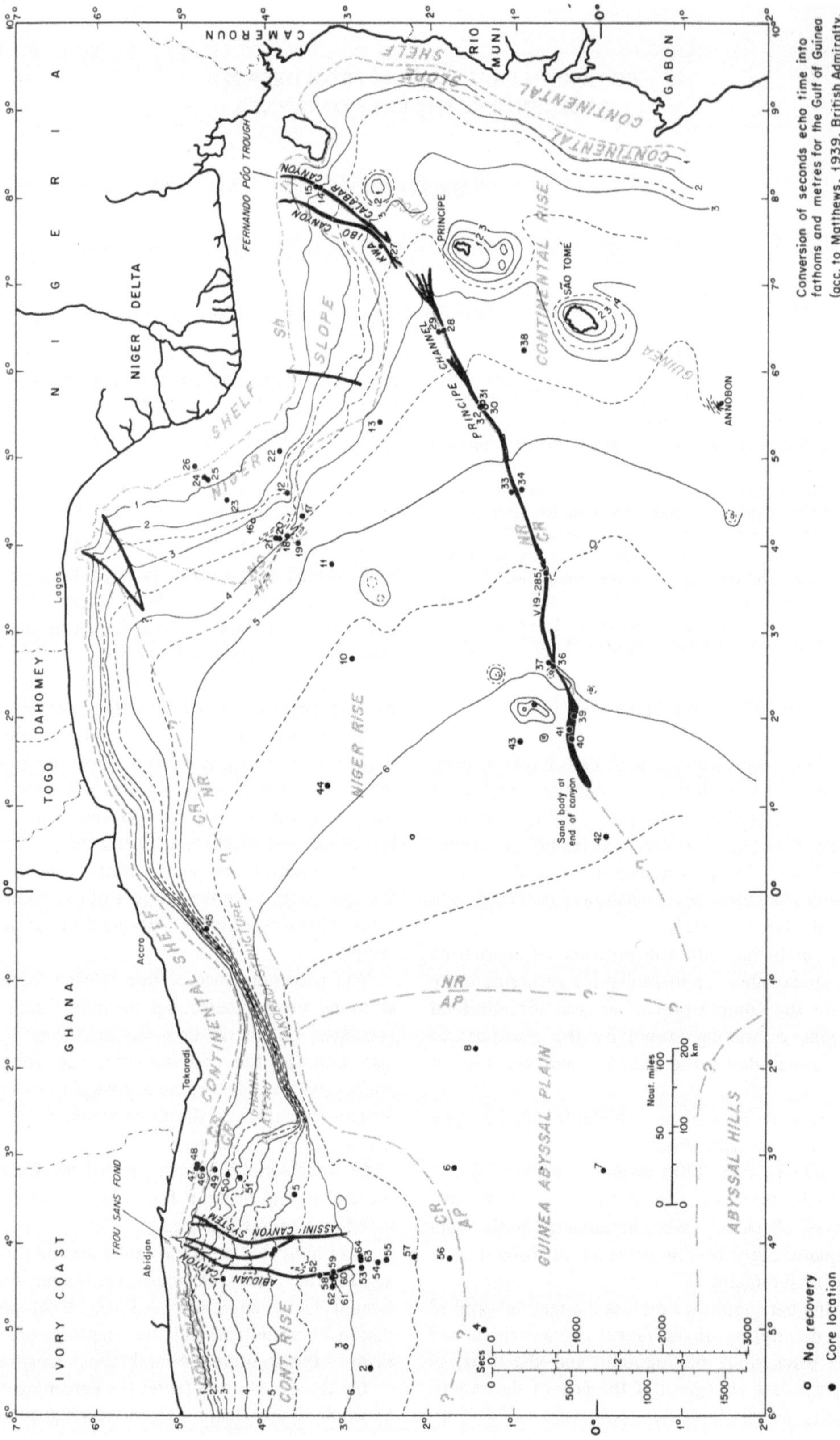

Fig. 1
Isobath chart with morphologic units and topographic names.

the percentages should total one hundred, but in most samples they do not, the remainder being undetermined. In those samples where the amount of specimens in the three fractions appeared to be too low for percentage counts, the occurrences are indicated by x.

The species determinations were accomplished with the aid of the literature indicated in the bibliography, while a check list of species is given in a separate appendix.

A number of quantitative faunal parameters have been assembled in tables 7-10, comprising the following items:

W % R	=	weight-percentage of the three residues combined versustthe dry weight of the sample
FN (foraminiferal number)	=	$\frac{\text{foraminiferal specimen counts fractions} > 150\ \mu \times \text{split factors}}{\text{dry weight of sample in grammes}}$
PFN (planktonic foraminiferal number)	=	$\frac{\text{planktonic specimen counts fractions} > 150\ \mu \times \text{split factors}}{\text{dry weight of sample in grammes}}$
BFN (benthonic foraminiferal number)	=	$\frac{\text{benthonic specimen counts fractions} > 150\ \mu \times \text{split factors}}{\text{dry weight of sample in grammes}}$
%P (planktonic/benthonic ratio)	=	$\frac{\text{PFN}}{\text{FN}} \times 100$
% shallow	=	percentage of specimens in total benthonic fauna having a known exclusively shallow-marine or non-marine distribution
% Pwu	=	combined percentage of characteteristic components of the *Planulina wuellerstorfi fauna* (see next chapter)
FGR (number of forams per gramme of residue)	=	$\frac{\text{foraminiferal specimen counts fractions} > 150\ \mu \times \text{split factor}}{\text{dry weight of residue} > 150\ \mu \text{ in grammes}}$
FN fractions > 75 μ	=	$\frac{\text{foraminiferal specimen counts fractions} > 75\ \mu \times \text{split factors}}{\text{dry weight of sample in grammes}}$

III. FAUNAL CHARACTERISTICS

A. *Foraminifera and sediment*

The core material, taken between roughly 300 and 5000 m sea depth, yielded in principle three types of benthonic fauna, dependent on environmental conditions on the sea floor itself, and two types of planktonic fauna, reflecting temperature fluctuations in the surface waters of the ocean.

Normally, the sea bottom of this part of the ocean consists of fine-grained material, but owing to the fact that quite a number of stations are situated in or near submarine canyons, coarse-grained material has often been found. The presence of larger amounts of coarse material will in this case be responsible for larger wash residues.

Tables 7-14, where the samples are grouped according to the sediment present in the residues, show that the samples with a shale residue have W%R values rarely exceeding 20, whereas the sandy residues, often have values exceeding 60. In tables 9 and 10 we also included for comparison three samples from a core taken in the Adriatic Sea (Brouwer, 1967).

Later on, it will be demonstrated that these large residues are related to submarine canyons and invariably contain high percentages of displaced shallow-water Foraminifera.

First, however, we will discuss the composition of the benthonic foraminiferal faunas.

B. *Benthonic Foraminifera*

1. Faunal types

As mentioned before, three main types of benthonic thanatocoenoses have been found in the sample material studied. These are

i a calcareous abyssal fauna,
ii as (i), but mixed with shallow-water components,
iii a calcareous bathyal – shallow abyssal fauna.

i. The deepest autochthous fauna present in the cores is the *Planulina wuellerstorfi* fauna, with the following characteristic species, of which at least two should be present as a main component (i.e. constituting 5% or more of the total (benthonic fauna):

Eggerella bradyi (Cushman)
Pullenia bulloides (d'Orbigny)
Globocassidulina subglobosa (Brady)
Epistominella exigua (Brady)
Oridorsalis umbonatus (Reuss)
Nuttallides umboniferus (Cushman)
Planulina wuellerstorfi (Schwager)
Melonis pompilioides (Williamson)

indicated in tables 1-6 with arrows.

This *Planulina wuellerstorfi* fauna has been found in all cores taken below 1350 m sea depth (see tables 7-10). The combined percentages of the eight components gradually increase with depth (textfig. 2), indicating a gradual transition into a shallower fauna group with decreasing depth.

At its deepest occurrence at 5084 m (the deepest core taken), it is not yet replaced by the *Rhabdammina* fauna (Brouwer, 1965), which occurs in the oceans of the world below the calcium-carbonate compensation depth, representing the deeper abyssal environment and the deepest marine benthonic assemblage known.

Traces of the *Rhabdammina* fauna are present in some samples, but the combined percentage of its components rarely exceeds 10%. The strongest influx is present in core 11 (5-8 cm), taken on the Niger Rise at about 3880 m sea depth, where *Adercotryma glomerata* (Brady), *Cyclammina trullisata* (Brady), *Hormosinidae, Rhabdammina abyssorum* Sars, *Glomospira charoides* (Jones & Parker) and *Trochammina globigeriniformis* Parker & Jones form 23% of the benthonic fauna.

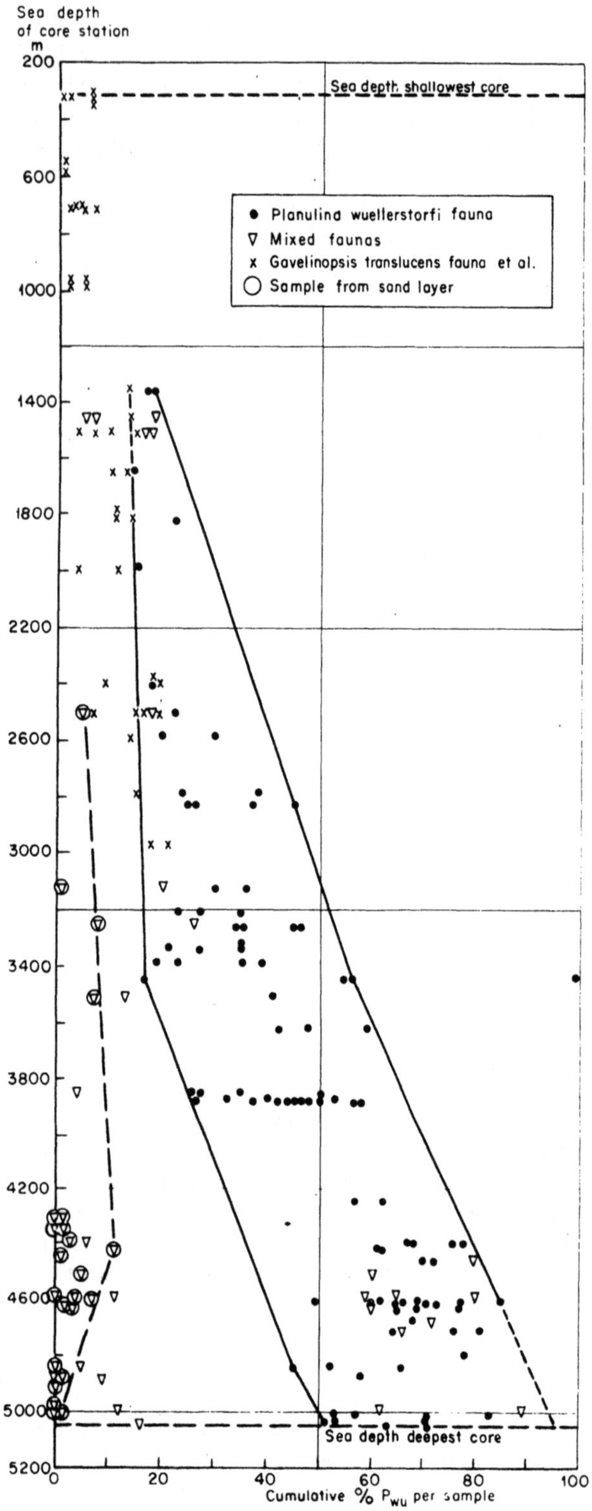

Fig. 2

ii. Quite a number of samples yielded faunas with variable percentages of forms, which are known to be characteristic of the inner- and middle-neritic environments (indicated in tables 1-6 as a separate group):

Ammonia beccarii (Linné)
Amphistegina lessonii d'Orbigny
Asterigerinata mamilla (Williamson)
Cancris oblongus (Williamson)
Cibicides ex gr. *lobatulus* (Walker & Jacob)
Cribroelphidium gunteri (Cole)
Cribrononion advenum (Cushman)
Cribrononion incertum (Williamson)
Dyocibicides sp. indet.
Elphidiidae
Elphidium crispum (Linné)
Eponides repandus (Fichtel & Moll)
Florilus asterizans (Fichtel & Moll)
Florilus atlanticus (Cushman)
Hanzawaia ex gr. *concentrica* (Cushman)
Miliolidae
Neoconorbina orbicularis (Terquem)
Nonionella sp. indet.
Pararotalia armata (d'Orbigny)
Planorbulina mediterranensis (d'Orbigny)
Poritextularia panamensis (Cushman)
Ptychomiliola separans (Brady)
Quinqueloculina
Reussella miocenica Cushman
Rosalina globularis d'Orbigny
Textulariidae

In particular in the samples with a sandy residue (see tables 11-12), this group can constitute up to 85% of the total benthonic fauna. In most samples these shallow components co-occur with the *Planulina wuellerstorfi* fauna; obviously the latter fauna is increasingly suppressed with an increasing percentage of shallow-water forms* (text fig. 3). Moreover, the very large residues with a W%R exceeding 80 always have a high percentage of shallow-water forms (text-fig. 4).

The cores containing samples with shallow-water components in a sandy residue are all taken from the Principe Channel and Abidjan Canyon, or from the submarine fan on the abyssal plain in front of the Abidjan Canyon (see textfig. 5). (See Houbolt, 1973, who describes these deep sea canyons in detail).

The presence of displaced faunas in abyssal sediments was recognised by various authors, including Phleger (1951), Phleger, Parker & Peirson (1953), Ericson et al. (1961) and Davies (1968). The latter concluded on the investigation of 221 Atlantic and Caribbean cores that

* From the shelf area N. of Fernando Poo, Berthois, Crosnier & Le Calvez (1968) mention the occurrence of a poor, in part badly preserved, shallow-water fauna with a number of components, identical to the displaced faunas. The percentage of displaced components may in reality be larger, as some species (e.g. ***Brizalina subaenariensis*** (Cushman) and ***Euuvigerina peregrina*** (Cushman) have a very wide depth distribution, and are actually found as occurring frequently or abundantly at water-depths of 200 m or less just W. of Abidjan (Le Calvez, 1963).

"anomalous layers of sand, silt, and lutite occur widely in the deep basins of the Atlantic. Evidence for deposition of these layers by turbidity currents is as follows: (1) the layers occur in submarine canyons, in deltalike features at the terminal ends of canyons, in basins and depressions, never on isolated rises; (2) they are interbedded with late Pleistocene sediments of abyssal facies; (3) they are well-sorted and commonly graded; and (4) they commonly contain organic remains of shallow-water origin."

Values of up to 78% shallow-water forms are also described by B a n d y (1964) in sand beds of the submarine fans of turbidites in the San Pedro Basin (approx. 1000 m deep), California.

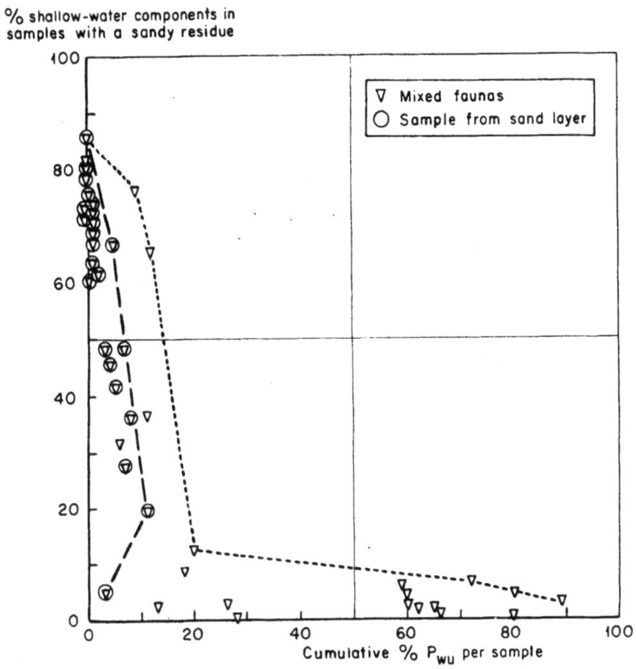

Fig. 3

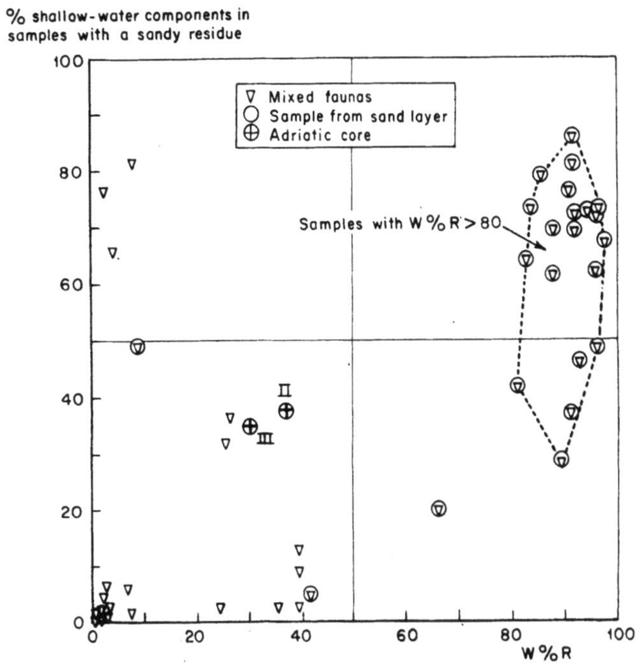

Fig. 4

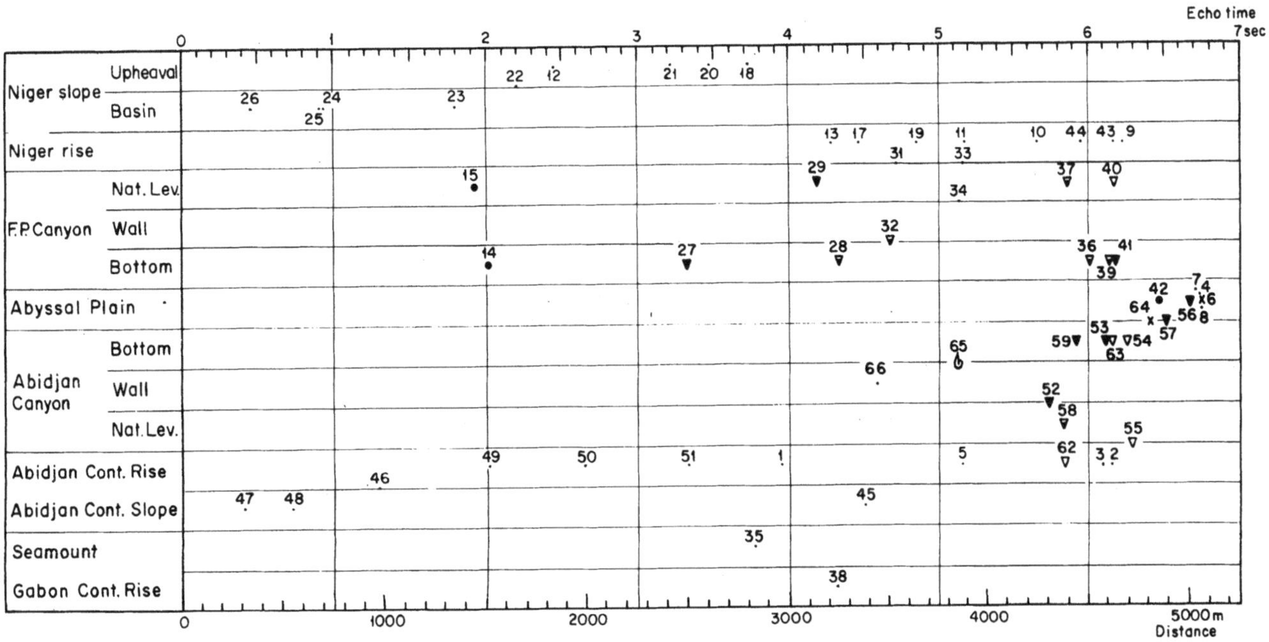

Fig. 5

Distribution of cores with sandy residues, with
∇ < 10% shallow-water forms } for each core the sample with
▼ 11-50% " " } the highest percentage is taken
▼ > 50% " "

• With a residue of plant material
x With almost barren samples of shale with shallow-water forms
ɓ With a residue of grey shales rich in Mollusca and presence of shallow-water forms

Summarising, we conclude that foraminiferal faunas, consisting for the greater part of components indicative of shallow neritic environments, may under certain circumstances occur in abyssal sediments.

Apart from the clearly displaced faunas just mentioned, some cores from the canyons or submarine fan differ from the normal pattern in other respects.

In cores 14, 15 and 42, some samples yielded residues consisting almost exclusively of displaced plant material, sometimes co-occurring with shallow-water Foraminifera.

From cores 6 (50-53 cm) and 64 (248-251 cm), hardly any residue remained after washing. These samples are almost barren, but do contain some shallow-water forms and could very well represent the almost barren upper interval of horizontal lamination "a" of a turbidite (compare B r o u w e r, 1967, p. 233, G r i g g s & F o w l e r, 1971).

In one sample of grey shale (core 65: 156-158 cm), the residue is rich in bioclastic material, together with some displaced forams.

The cores with intervals in which these aberrant characteristics have been observed are indicated in textfig. 5.

iii. The third main type of fauna is formed by the *Gavelinopsis translucens* fauna, with

Bulimina alazanensis C u s h m a n
Bolivina albatrossi C u s h m a n
Gavelinopsis translucens (P h l e g e r & P a r k e r)

which have been found as a main component in deeper bathyal and shallow abyssal faunas in the Gulf of Mexico.

In the Gulf of Guinea this group is not very well developed; the combined percentage of these three species does not exceed 19%. One of the three species is present as a main component in core stations 21, 46, 49, 50 and 51, between 980 and 2500 m sea depth.

At 2830 m (core 35) this fauna already occurs together with the *Planulina wuellerstorfi* fauna, making the boundary between these two faunas of a transitional character.

In this third group of faunas, we have also included other samples not bearing a *Gavelinopsis translucens* fauna; they are often characterised by *Sphaeroidina bulloides* d'Orbigny, *Hoeglundina elegans* (d'Orbigny), *Osangularia culter* (Parker & Jones) and other species of a worldwide bathyal and shallow abyssal distribution.

The samples considered as having faunas of the third group are listed in tables 13 and 14.

2. Statistical parameters

The number of benthonic Foraminifera $> 150\,\mu$ per gramme of sediment (BFN) is rather low for the *Planulina wuellerstorfi* fauna (text fig. 6), generally between 1 and 100. For the *Gavelinopsis translucens* fauna et al., most values lie between 30 and 300, whereas the mixed faunas show a spread between 1 and 300.

The distribution of BFN against depth* (text fig. 7) shows a weak trend of decreasing values with increasing depth, except for the mixed faunas.

The correlation between benthonic fauna and lithology is demonstrated in text figure 8. Both autochthonous faunas, the *Planulina wuellerstorfi* and the *Gavelinopsis translucens* fauna, are practically confined to a muddy bottom, whereas the displaced faunas occur with higher concentrations in sands and with lower concentrations in muds.

W a l t o n (1964) remarked that the shallower the environment, the smaller the number of species and the higher the percentage dominance of the dominant species. His species counts are based on residues $> 62\,\mu$ and his data originate from stations generally taken shallower than 400 m sea depth.

This relation, however, seems to be reversed again at greater depths, as in some of our samples taken at depths of about 5000 m, we found in the residues $> 70\,\mu$ a low number of species combined with a high percentage dominance. According to Walton (his fig. 26), this would indicate a sea depth of less than 10 fathoms.

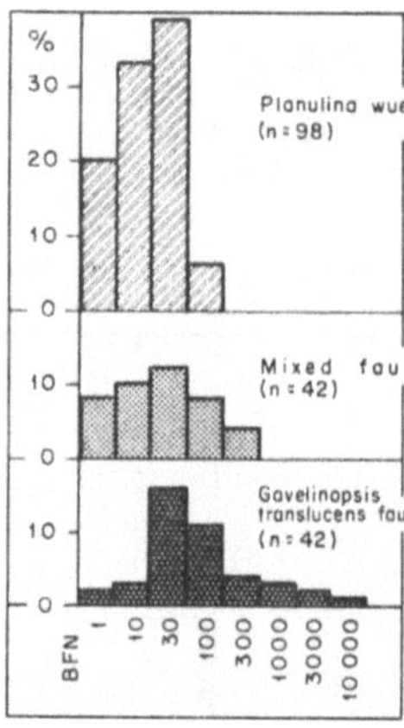

Fig. 6

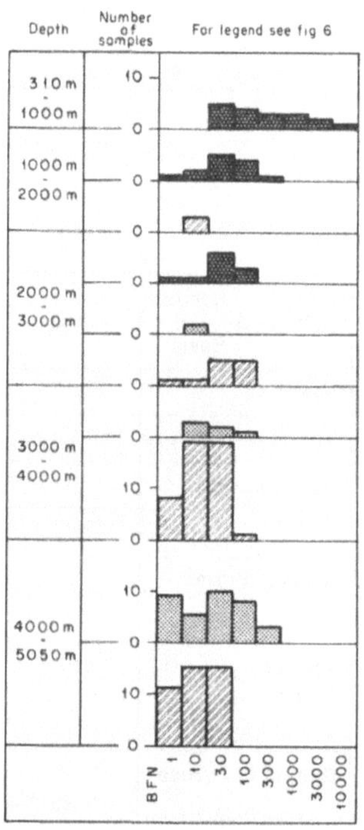

Fig. 7

C. *Planktonic Foraminifera*

1. Faunal types

The top part of many cores contains a planktonic fauna identical to the one living at present in this part of the

* Although most samples are not strictly Recent, but Holocene and Pleistocene, the difference in sea level is considered negligible if arranged in depth groups of 1000 m.

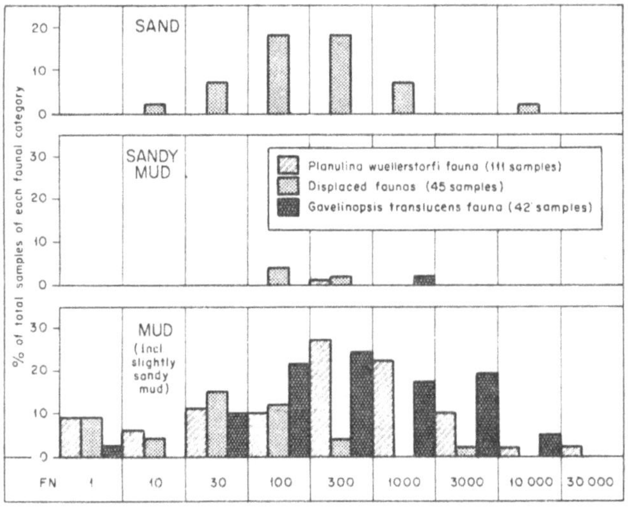

Fig. 8

Atlantic Ocean (S c h o t t, 1937, p. 99; P h l e g e r, P a r k e r & P e i r s o n, 1953, p. 107), with

Globigerinoides conglobatus (B r a d y)
Globigerinoides ruber (d'O r b i g n y)
Globigerinoides sacculifer (B r a d y)
Globorotalia menardii (d'O r b i g n y)
Orbulina universa d'O r b i g n y
Pulleniatina obliquiloculata (P a r k e r & J o n e s)
Sphaeroidinella dehiscens (P a r k e r & J o n e s)

In their figure 13, Phleger et al. indicate *Globorotalia menardii* as occurring in percentages of over 10% only southwards of 20° NL. This is more or less in agreement with B é & H a m l i n (1967, text fig. 35) who show a patchy distribution of up to 5% between 20° and 40° NL for this species. On this basis we consider those samples containing a planktonic fauna with 10% or more *Globorotalia menardii* as reflecting the same climatic conditions in the surface waters as those prevailing nowadays in the Gulf of Guinea, which is geographically located between the equator and 5° NL.

At present, *Globigerina pachyderma* (Ehrenberg) and *Globigerina bulloides* (d'Orbigny) have their main distribution northwards of 40° NL (B é & H a m l i n, 1967), although P h l e g e r, P a r k e r & P e i r s o n (1953, fig. 3) did record some high percentages of *G. bulloides* as far south as the equator. *Globigerina inflata* (d'Orbigny) also indicates colder surface water, but has a slightly more southward extension than the other two species.

Samples containing these three species with a combined percentage of 10% or more and in which *Globorotalia menardii* is either absent or present only in low percentages, have been interpreted as being deposited during a time when the surface waters were colder than they are today.

Some samples contain a mixed cold/warm planktonic fauna, such as those mentioned by P h l e g e r, P a r k e r & P e i r s o n (1953, p. 108).

2. Age determination

The top layer with abundant *Globorotalia menardii* (mentioned in the previous paragraph as reflecting present-day climatic conditions), is considered as having been deposited during Recent/Holocene times (see S c h o t t, 1937; E r i c s o n et al., 1961, 1963, 1964, and others).

Underneath this top layer, the planktonic faunas often indicate conditions colder than they are at present. This interval is determined as being of Pleistocene age, in accordance with the same authors.

In some cores *G. menardii*-rich layers occur, underlying the cold-water intervals. These layers have planktonic faunas which have essentially the same specific composition as those from the Recent/Holocene, but which differ in that they are often characterised by the presence of abundant *Globorotalia tumida* (Brady), considered by some authors as a variant of *G. menardii*.

Such samples are interpreted as having been derived from an interglacial stage of the Pleistocene (E r i c s o n et al., 1961, p. 263, zone X; p. 269, zones X and V).

In the present material, *Globorotalia menardii* and variants are predominantly sinistrally coiled in all samples, except for those from the deeper part of core 5 where they are dextrally coiled. This implies that pre-Pleistocene beds have been penetrated in this core (E r i c s o n et al., 1963; B e r g g r e n, 1968, p. 296).

One sample from this lower interval has been studied in more detail, resulting in an assemblage of species (see textfig. 9) characteristic of the *Globoquadrina altispira* zone of Pliocene age (P o s t u m a, 1971).

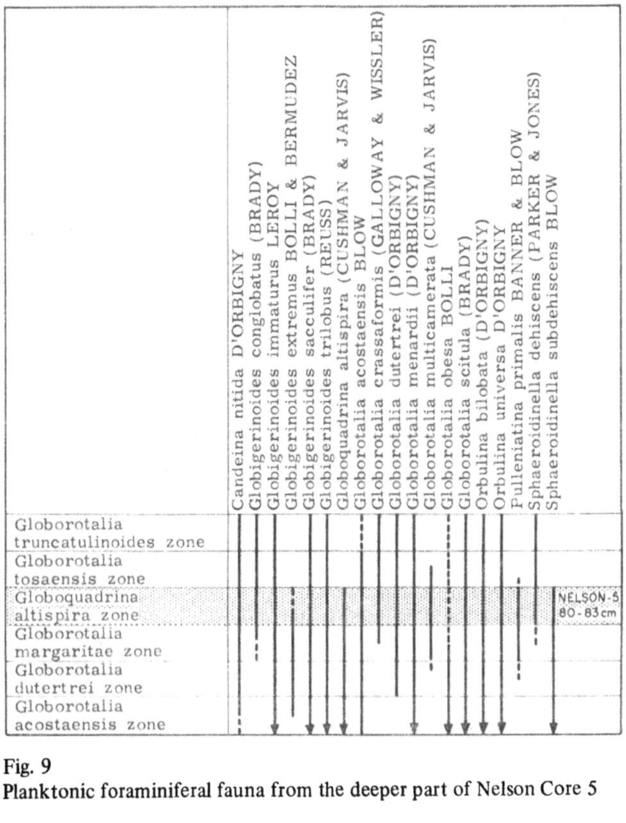

Fig. 9
Planktonic foraminiferal fauna from the deeper part of Nelson Core 5

3. Statistical parameters

The relation between %P and depth is expressed in text fig. 10.

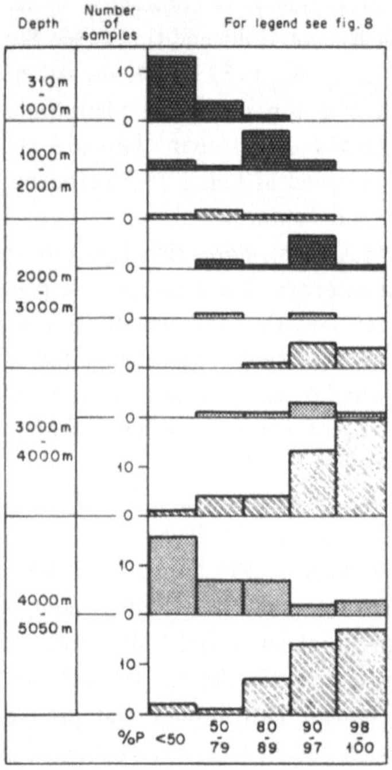

Fig. 10

The *Planulina wuellerstorfi* fauna occurs mainly in samples with over 90% planktonics. This trend is disturbed in the mixed samples, while in the samples with a *Gavelinopsis translucens* fauna et al., the percentage of planktonics decreases towards shallower depths.

In a general way this trend is in accordance with Grimsdale & van Morkhoven (1955), who apply this value as an indicator of sea depth (see also Funnell, 1967, p. 339-340).

If compared with the percentage of shallow-water components (text fig. 11), the values of %P Show a trend of decreasing %P with an increasing percentage of shallow components, in fact, a "dilution" effect.

A comparison of %P with PFN (text fig. 12) indicates for the *Planulina wuellerstorfi* fauna a narrowing-down within the 95-100%P limit for the PFN values over 1000, and even to the 99-100% limit for the values over 10 000.

The mixed and *Gavelinopsis translucens* faunas behave quite differently. For the two autochthonous faunas, the PFN values range roughly between 30 and 10 000, but for the mixed faunas the values are on average lower (text fig. 13 & 14).

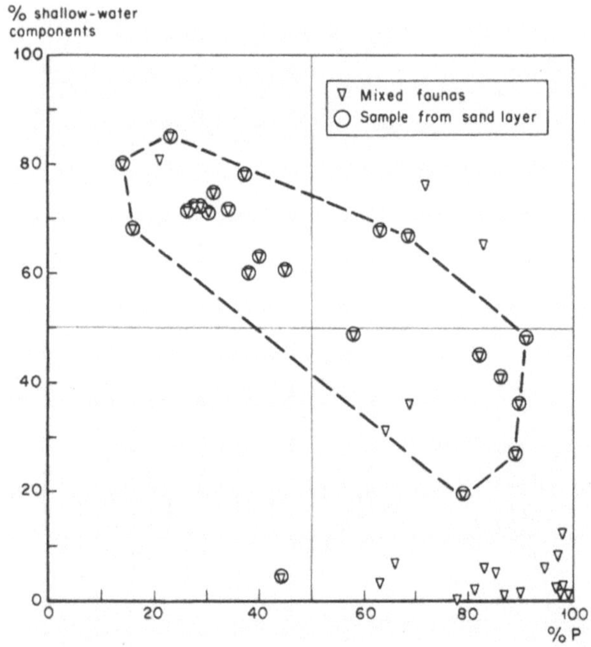

Fig. 11

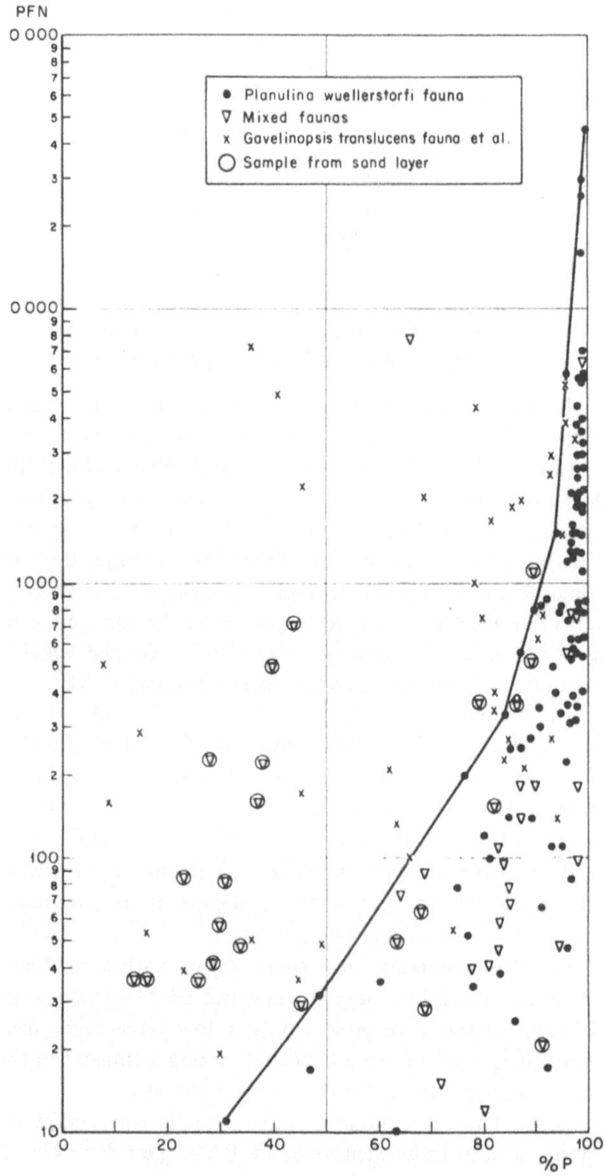

Fig. 12

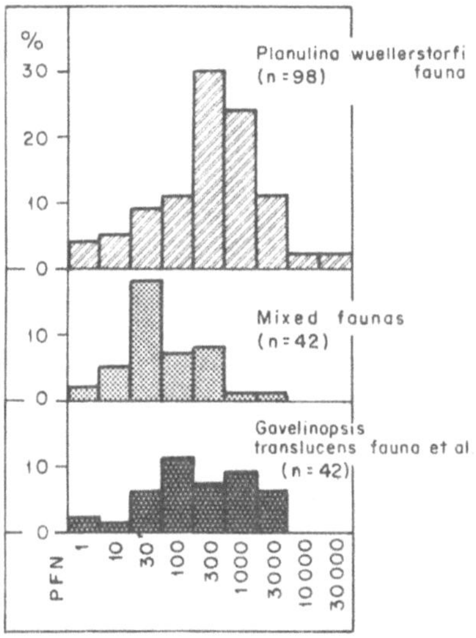

Fig. 13

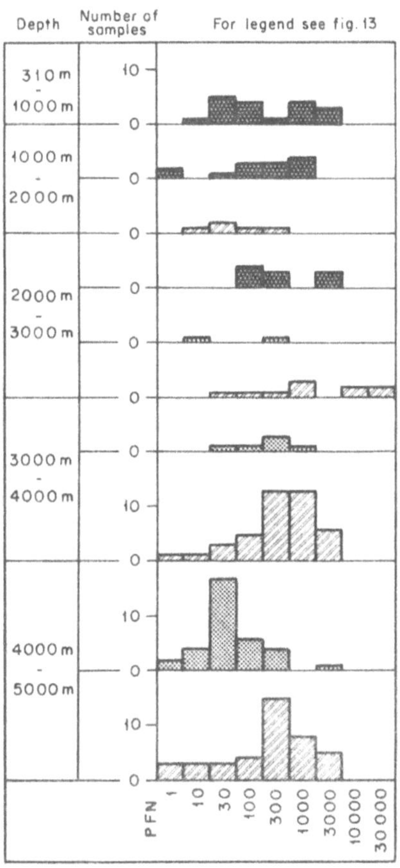

Fig. 14

D. *Rate of sedimentation*

The rate at which sediment is deposited depends on a number of factors (sediment supply, nature of sediment, water movement, etc.), and varies from very fast sedimentation to virtually no sedimentation at all, not considering erosion.

In areas with no supply of anorganic sediment, the only deposition to occur is that formed by the remains of organisms living in the overlying water column or on the sea floor itself. The more the number of foraminiferal tests present in one gramme of sediment approaches the number of tests contained in one gramme of Foraminifera, the slower the rate of deposition will be.

The foraminiferal number (FN) was introduced by S c h o t t (1937, pp. 112-113 under the name "Foraminiferenanzahl", a number representing the quantity of Foraminifera in the fraction 200-2000 μ in one gramme of dried sediment, obtained not by counting, but by weighing the wash residue, as follows:

> "Die gröbste Schlämmfraktion der Grundproben (D = 2,0-0,2 mm) enthält meist ohne andere Gemengteile vorwiegend die Foraminiferenschalen; es kann somit aus dem Gewicht dieser Fraktion und aus der Einwage der zum Schlämmen benutzten Grundprobe gewichtsmäszig die Foraminiferenanzahl in einem Gramm getrockneter Probe bestimmt werden, da nach einer Auszählung 10 000 ausgewachsene Foraminiferen durchschnittlicher Grösze 0,4288 g wiegen. (Zur Bestimmung dieses Gewichtes sind 4000 Foraminiferen ausgezählt und gewogen worden.) Die so gewonnenen Zahlen geben natürlich nicht die wahre Anzahl der unzerbrochenen Foraminiferenschalen in den einzelnen Proben wieder; denn bei dieser gewichtsmäszigen Bestimmung konnte das Gewicht der Foraminiferenbruchstücke, die in der Fraktion D = 2,0-0,2 mm vorhanden sind, von den unzerbrochenen Foraminiferenschalen nicht getrennt werden. Auszerdem enthalten die feineren Fraktionen kleinere Individuen, die gleichfalls nicht berücksichtigt wurden.'

Using the above figure obtained by Schott from plankton-rich sample material taken in the Mid Atlantic, we arrive at a value of FN = 23 320 for a pure *Globigerina* ooze.

S a i d o v a (1961) found values up to 120 000 for pure foraminiferal oozes present on the submarine Hawaii ridge, but if these values, based on the number of specimens present in the fractions $> 50\ \mu$, are compared with a nomogram compiled by B e r g e r (1969, fig. 2B), they are about ten times too low with respect to the value calculated for the $> 200\ \mu$ fractions. This discrepancy can be explained in part by the variability in density and medium size of foraminiferal tests in different geographical and environmental areas.

In order to obtain some calibration figure for the Gulf of Guinea samples, a number of counts have been made on material from core 35 (246-253 cm). The residue of this sample consists entirely of foraminiferal tests. In three 1/1000 g portions of the fraction 150-420 μ, a total of 66, 76 and 80 specimens were counted; in three similar portions of the fraction $> 420\ \mu$, the countings resulted in 27, 28 and 35.

These figures lead to an average of 64 000 (± 7000)

specimens in the size fractions > 150 μ for one gramme of Foraminifera. This figure agrees very well with Schott's values, if compared with Berger's nomogram.

C o r r e n s (1937) suggested that the term *Globigerina* ooze be restricted to samples with Fn = > 6000. This figure, compared with Schott's value of 23 320 for a pure *Globigerina* ooze, would indicate that at least one quarter of the sediment consists of unbroken foraminiferal tests. Such samples would contain over 60% $CaCO_3$ (C o r r e n s, 1939, fig. 68).

Our FN values are based on the fractions > 150 μ, although the 75-150 μ fraction has also been studied. The lower limit of 150 μ is taken, however, to restrict the degree of error, which increases as the ratio split factor/countings becomes higher.

Compared with our average of 64 000 specimens for a pure *Globigerina* ooze, the lower limit of such an ooze would be represented by a value FN = 25 000.

Those values have only been found in the sea-mount samples from core 35 (text fig. 15); thus *Globigerina* ooze has been deposited at this place. For the pelagic sediments containing a *Planulinaa wuellerstorfi* or *Gavelinopsis translucens* fauna et al. the rate of sedimentation is higher as the bulk of the sediment is formed by mud (compare also text fig. 8).

The same trend appears if the ratio FGR/FN is plotted graphically (text fig. 16). The pelagic sediment samples show ratios between 5 and 1000, depending on the amount of residue > 75 μ left after washing. The sea-mount samples with a *Globigerina* ooze, with a large residue > 75 μ, and also the sand with displaced faunas, also giving a large residue > 75 μ, consequently have ratios approaching 1.

The *actual* rate of sedimentation can be calculated from the sediment thickness divided by the deposition time.

For the Atlantic Ocean, the following values have been chosen from the literature:

Equatorial Atlantic	Rate of accumulation mm/1000 yr	Source
Globigerina ooze	12	S c h o t t, 1937
Red deep-sea clay	<8.6	
Average for deposits between 3330 and 4515 m	33	E r i c s o n, 1963
Pelagic sediments, general	10-66	G r i f f i n et al., 1968
Idem, Mid-Atlantic Ridge	2-7	

Taking the end of the Pleistocene at around 11 000 years B.P. (E r i c s o n, et al., 1964), we arrive at the following estimates for the Holocene sediments in the Gulf of Guinea:

Niger slope (basin)	50-100 mm/1000 yr
(upheaval)	20–50
Niger rise	0-60
Fernando Poo Canyon	0-120
Gabon Sea mount	10
Abidjan slope	0
Abidjan rise	0-50
Abidjan canyon	0-80
Abidjan abyssal plain	10-30

These estimates indicate that the rate of sedimentation off Abidjan is on average slower than that on the Niger slope. On the Gabon Sea Mount the rate is very slow, as already implied by the presence of *Globigerina* ooze. The rates found for the abyssal plain coincide in general with those for pelagic sediments.

For sediments deposited by turbidity currents, these rates can, of course, become much higher and much more variable. From the radiocarbon dating of six cores taken in the Santa Monica – San Pedro Basins along the Californian coast, with depths not exceeding 100 m, E m e r y, H ü l s e m a n n & R o d o l f o (1962) calculated an average rate of 700 mm/1000 yr for such sediments.

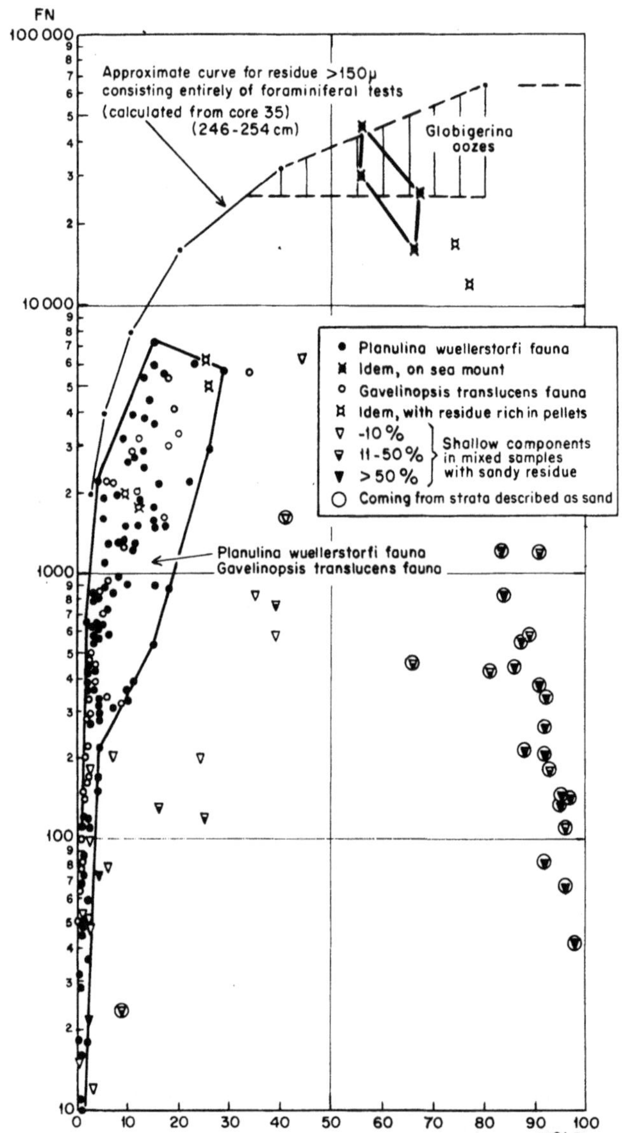

Fig. 15

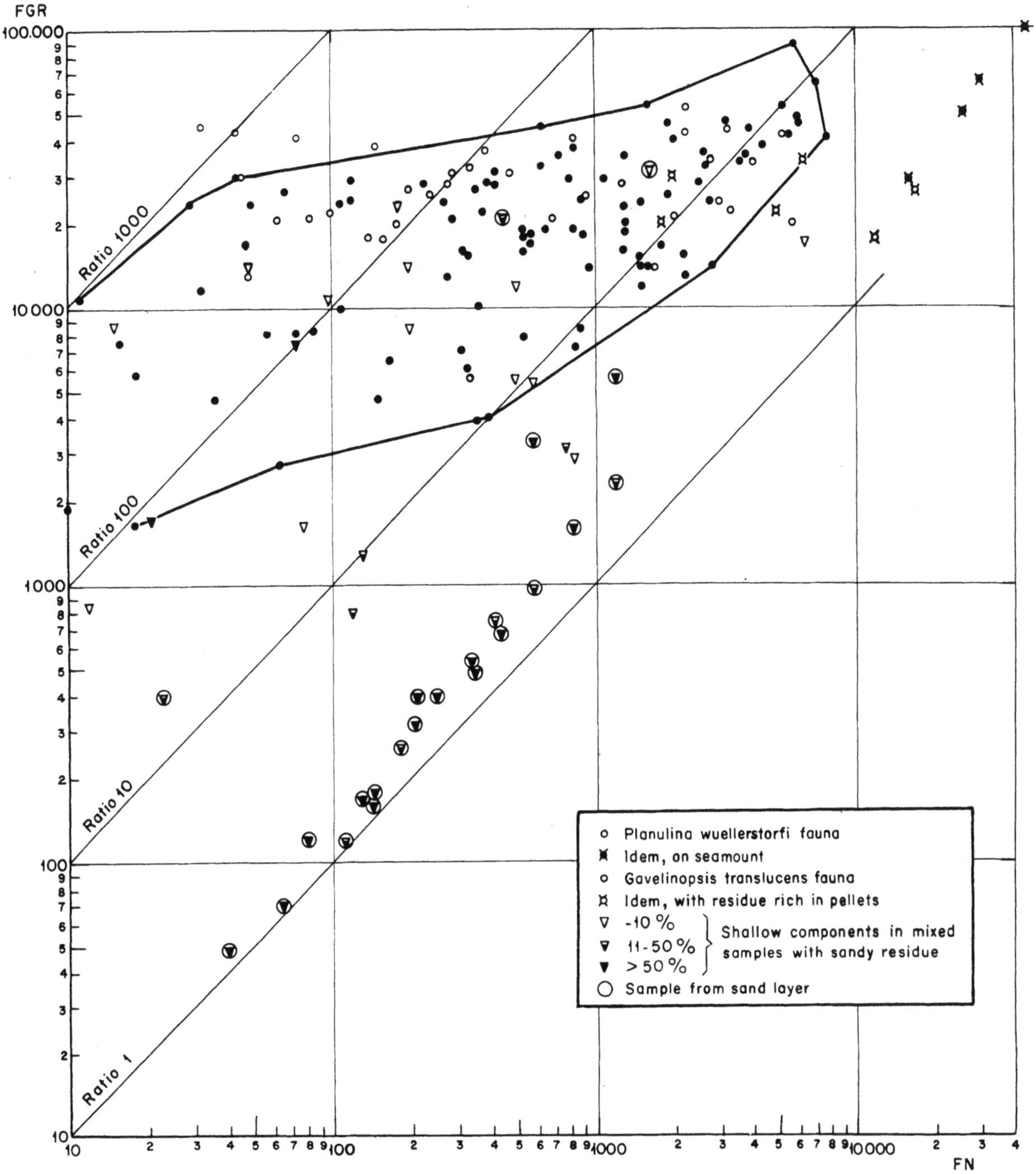

Fig. 16

As none of the cores with displaced sediments penetrated a palaeontologically measurable part of the Pleistocene, it is unfortunately impossible to give values for the rate of sedimentation in the canyon sediments containing sandy deposits.

E. *Solution at depth*

From the distribution charts, it appears that the composition of the planktonic faunas varies considerably from sample to sample and from station to station.

As already pointed out, climatic changes in the surface water are reflected in the specific composition of the planktonic fauna, but this does not explain why some *Globorotalia menardii* faunas contain abundant *Globigerinoides ruber* and others do not.

The reason for this can be found in the differential solution of planktonic species at greater oceanic depths. This phenomenon has been studied in more detail by Ruddiman & Heezen (1967), who conclude that

"In deeper realms such variations are largely a product of differential solution among the species. This leads to preferential enrichment of resistant, deep dwelling species *(Globorotalia tumida, Sphaeroidinella dehiscens, Globigerina eggeri,* and *Pulleniatina obliquiloculata)* and to selective elimination of fragile, surface-inhabiting individuals *(Globigerinoides ruber).* The shallowest depth at which differential solution is effective in Equatorial Atlantic sediments is at present 4000 m. Most faunal assem-

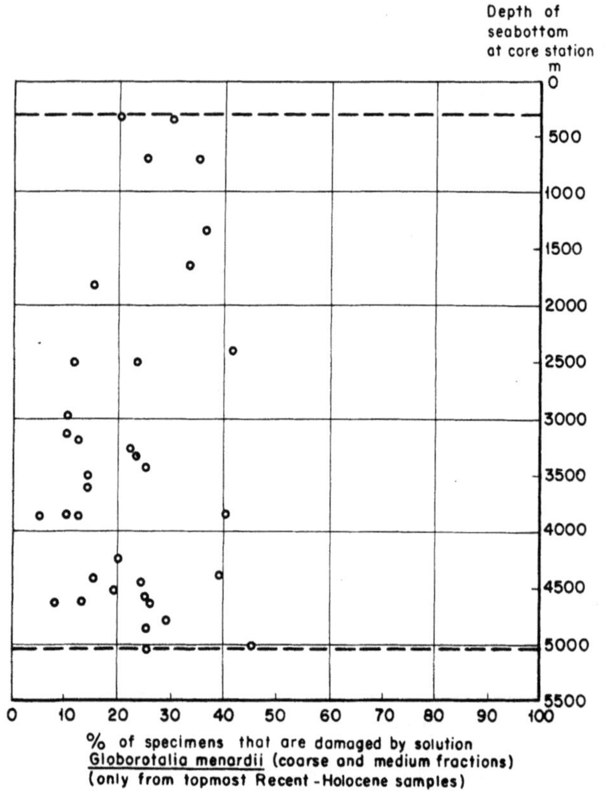

Fig. 17

blages in depths greater than 4500 m are differentially dissolved, and in depths greater than 5000 m all show severe effects."

Their data also suggest that *G. menardii* is little affected by differential solution on a percentage basis, which would make this species a good marker for tropical faunas.

The distribution charts clearly demonstrate high percentages of *G. ruber* in shallower stations, while in the deeper cores this species is often completely missing in the *G. menardii* faunas.

As already pointed out by Phleger, Parker & Peirson (1953, p. 117 seq., plate 12), the effect of solution is also visible on the tests of Foraminifera in the form of holes and solution breaks. For a number of Recent-Holocene samples, we calculated the percentage of *G. menardii* showing these signs, and plotted the obtained values against depth (text fig. 17). However, no trend becomes apparent, which is not surprising as, on the basis of both laboratory experiments and thermodynamic theory, and as actually proven experimentally in the ocean (Berger, 1967; Hudson, 1967), solution is already noticeable at the lysocline, the level below which solution greatly increases (Berger, 1970: Parker & Berger, 1971). This level is not parallel with and occurs above the compensation depth below which essentially all calcareous matter is destroyed. According to a graph given by Revelle (1944), the compensation depth for the Atlantic is about 5500-6000 m, which is far below the depth at which the Gulf of Guinea cores were taken.

From the foregoing, the practical conclusion can be drawn that in fossil faunas equivalent to the *Planulina wuellerstorfi* fauna or the *Rhabdammina* fauna, which occur in Recent deposits right above or below the compensation depth, it is highly probable that the planktonic element is either incomplete by selective solution or completely absent.

F. *Other Organisms*

1. Radiolaria

The environmental relation between planktonic Foraminifera and Radiolaria is mentioned by Cifelli & Sachs (1966), who, from plankton samples taken in the Western Atlantic, come to the tentative conclusion that in surface waters the standing crop of Radiolaria is roughly comparable in number to that of planktonic Foraminifera.

In many samples from the Gulf of Guinea, particularly those containing the *Planulina wuellerstorfi* fauna, spumellarid and nassellarid Radiolaria (compare Riedel, 1963, fig. 7) are present in quantities of up to 40 000 per gramme of sediment in the fraction $> 75\ \mu$. Values of 10 000 and over are only found in samples with more than 90% planktonics.

However, none of the Gulf of Guinea samples has quantities approaching those given by Riedel (1951) from the Indian Ocean, i.e. 300 000-1 000 000 in the size fraction $> 50\ \mu$ from six localities between the Seychelles and Cape Guardafui in the Somali basin between 4065 and 5100 m sea depth.

On the other hand, Bandy (1961) found a maximum of 5200 for the size fraction $> 61\ \mu$ in the Carmen-Farallon Basin of the Gulf of California, which is much lower than that found in the Gulf of Guinea.

Only the high values from the Indian Ocean seem to approach those of a genuine radiolarian ooze, which up till now has only been recorded from a wide strip in the Pacific Ocean underlying the average path of the counter-equatorial current and several other isolated patches. It is apparently absent in the Atlantic (Piggot, 1944, fig. 1).

We are therefore inclined to relate the majority of the radiolarian cherts and other biogenic siliceous sediments from Miocene and older strata mentioned by Grunau (1965) and Ramsay (1971) to the radiolarian ooze from the Pacific, where it co-occurs with the (siliceous) Rhabdammina fauna, sponge spicules and manganese nodules (Murray & Renard, 1891).

2. Echinoidea

In numerous samples, spines and ambulacral plates of echinoids are present, sometimes in considerable numbers. Although common in the *Planulina wuellerstorfi* fauna, they attain higher concentrations in samples having high percentages of displaced shallow-water Foraminifera. There is no apparent correlation between depth and quantity.

3. Mollusca

Mollusca (small Gastropoda and Lamellibranchia, and planktonic Pteropoda) occur in many samples. In general they are concentrated in samples with displaced Foraminifera.

In sediments containing autochthonous foraminiferal assemblages, the amount of mollusks decreases irregularly with increasing depth.

Pure pteropod oozes have not been found in the present material. According to Bruun (1953), they are "of minor importance; even in the Atlantic, from where the purest typical deposits are known, they form only a small percentage of the total area". According to the same author they are occasionally found as deep as about 3500 m, but "the contribution of pteropods (and heteropods, which have a somewhat similar shell) to pelagic sediments is severely limited by the instability of their aragonitic shells" (Riedel, 1963, p. 873).

4. Ostracoda

Ostracods are rare, except in samples with displaced shallow material, where they attain concentrations of 50 to 600 valves per gramme of sediment.

5. Bryozoa

Bryozoa are only present in samples with more than 10% shallow-water foraminiferal components. All species are of a nearshore origin (Lagaaij, 1972).

6. Pisces

Neither fish teeth nor otoliths are common anywhere in the material studied.

IV. DISTRIBUTION OF RECENT AND FOSSIL PLANULINA WUELLERSTORFI FAUNA

The abyssal *Planulina wuellerstorfi* fauna is known to occur in the Recent deposits of the following areas:

Gulf of Guinea (abyssal, between 1350 and 5050 m sea depth) – this report.
North Atlantic (abyssal, 2105-5532 m) – Schott, 1935: Phleger, Parker & Peirson, 1953.
Gulf of Mexico (abyssal, 1280-3367 m) – Phleger & Parker, 1951; Parker, 1954.
Greenland Sea (abyssal, 1140-3000 m) – Stschedrina, 1947.
Weddell Sea (abyssal, 2198-4477 m) – Earland, 1936.
N.W. Pacific (abyssal, 2760-3050 m) – Saidova, 1961] co-occurring with *Rhabdammina* fauna.
N.E. Pacific (abyssal, 2440 m) – Saidova, 1964] co-occurring with *Rhabdammina* fauna.
Arctic (deep bathyal – abyssal), 740-2205 m) – Green, 1960.
Andaman basin (abyssal, 1800-3420 m) – Frerichs, 1967.

It has also been found in Pleistocene sediments of the:
Gulf of Guinea – this report.
North Atlantic – Schott, 1935; Phleger, Parker & Peirson, 1953.
Gulf of Mexico – Phleger & Parker, 1951; Phleger, 1955.

The fauna appears to be widespread in the Pliocene/Miocene, since it has been recorded from:
Colombia (Oligo-Miocene of the Carmen-Zambrano area) – Petters & Sarmiénto, 1956. Doubtfully Oligocene at some localities.
Czechoslovakia (Upper Silesian Basin) – Alexandrowicz, 1963.
Jugoslavia (Tortonian of Beograd, Stubik) – Petrovic, 1961, 1962.
Italy (Serie di Mombisaggio, Cantoni) – Ascoli, 1958.
Sicily (Flysch di Moleta, Caiazzo, with *Rhabdammina* faunas) – Ogniben, 1958.
Kar Nicobar (late Tertiary) – Frerichs, 1971.
Indonesia (late Tertiary of Sumatra, Kalimantan, Java, Buton) – Leroy, 1941a, b; Keyzer, 1953; Frerichs, 1971 and private information.
Timur – Tavares Rocha & Ubaldo, 1964.
Taiwan (Taitung) – Huang, 1964.
Gulf of Guinea (core station 5) – present report.

The areas and localities mentioned above are indicated on a world map text fig. 18), demonstrating the cosmopolitan nature of this fauna.

The number of occurrences is undoubtedly larger, but as many authors do not give the quantitative composition of the faunas they decribe, it is difficult to interpret their data with any degree of certainty.

An exception is made for Brady (1884, p. 662) who remarks that "*Truncatulina wuellerstorfi* is a common constituent of the deep-water ooze of all the great oceans". Although he does not specify the quantitative faunal composition for each station either, we have indicated on the world map those Challenger stations which probably contain this fauna.

V. RESULTS AND CONCLUSIONS

The environment, age and lithology of the sediments found in the Nelson cores are graphically presented on a sample basis in columnar sections of each core, scale 1 : 18 (fig. 19-22).

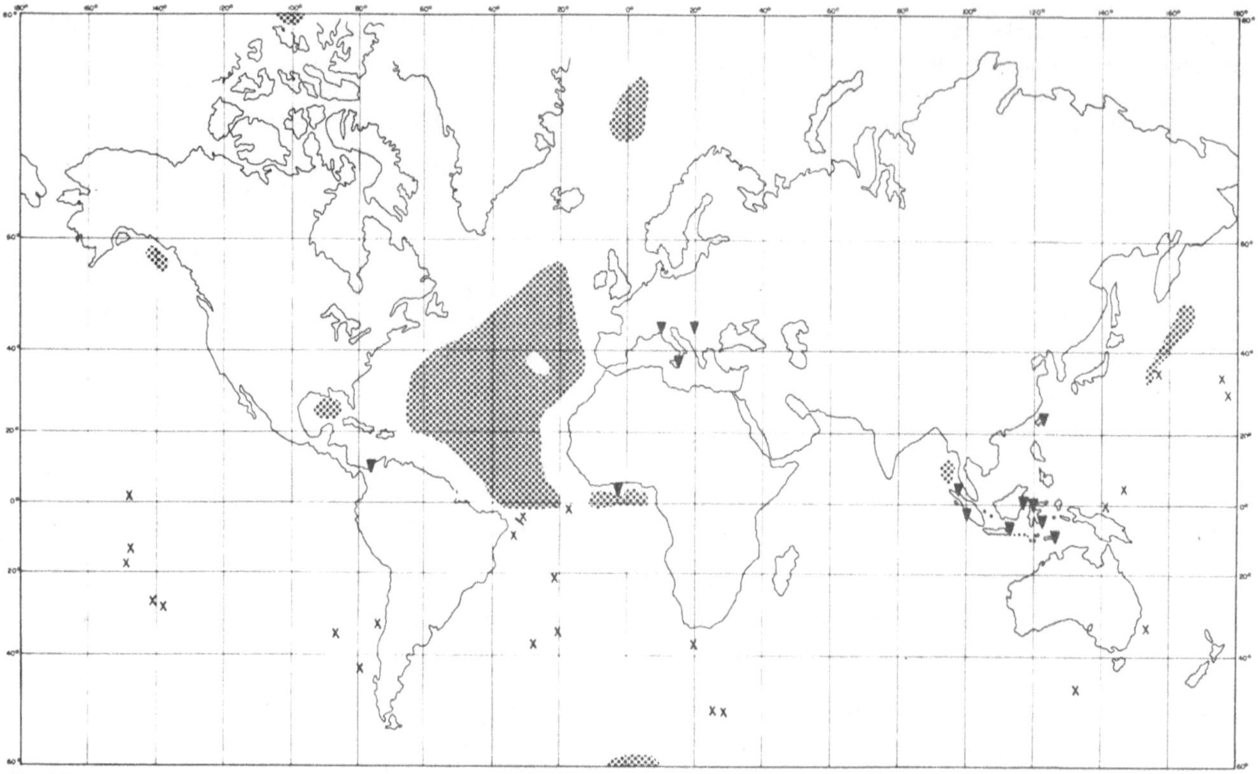

Fig. 18 Planulina wuellerstorfi fauna
Recent - Pleistocene (x Challenger stations)
▼ Miocene - Pliocene

Summarising, the following conclusions can be drawn:

i. The fine-grained pelagic sediments below 1350 m sea depth contain a benthonic assemblage, called *Planulina wuellerstorfi* fauna, with between 1 and 100 benthonic specimens in the size fraction $> 150\ \mu$ per gramme of total sediment. This fauna is usually accompanied by planktonic Foraminifera, which form over 90% of the total foraminiferal content. The sediments are deposited at a rate not exceeding 30 mm/1000 yr. The fauna has a world-wide distribution in Recent abyssal environments above the calcium carbonate compensation level, and has been noted in Neogene sediments from all over the world.

ii. *Globigerina* oozes are found on a sea mount with a benthonic fauna identical to that discussed in i.). The rate of deposition is estimated at about 10 mm/1000 yr.

iii. The fine-grained sediments below 300 m sea depth bear a benthonic *Gavelinopsis translucens* fauna et al., with between 30 and 300 benthonic specimens in the size fraction $> 150\ \mu$ per gramme of total sediment. Between 300 and 1000 m, it co-occurs with a planktonic fauna which forms less than 50% of the total foraminiferal components, increasing with depth to over 90% below 2000 m sea depth. The rate of sedimentation may attain 100 mm/1000 yr.

iv. The submarine canyons and fan deposits at the canyon head often contain sediments with high percentages of sand and displaced shallow-water Foraminifera in a "diluted" *Planulina wuellerstorfi* fauna. Values for the rate of sedimentation cannot be given. The presence of up to 85% inner-and middle-neritic benthonic elements in abyssal sediments is of practical significance for environmental interpretation in fossil strata.

v. The planktonic Foraminifera indicate the presence of a thin Recent/Holocene blanket of sediments overlying Pleistocene deposits in the Gulf of Guinea. One core reached strata of Pliocene age.

vi. Planktonic faunas, co-occurring with *Planulina wuellerstorfi* fauna and being close to the compensation depth, might not reflect the total planktonic fauna of the surface waters, as selective solution may eliminate thin-shelled components. Age determinations based on such faunas should be carried out with this in mind.

vii. Off Ibadan, the rate of sedimentation is slower on average than that on the Niger slope.

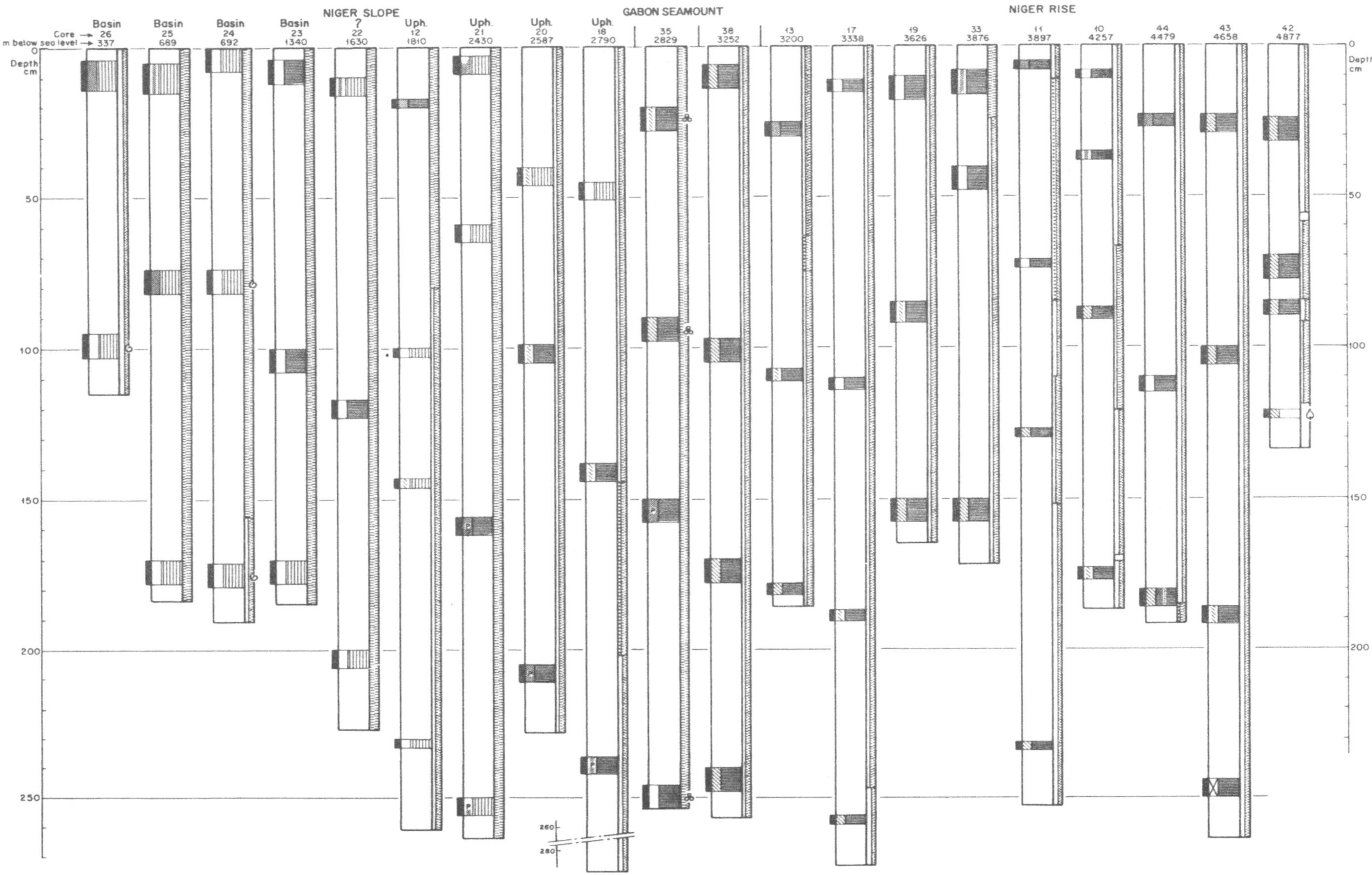

Fig. 19
Niger slope, Niger rise and Gabon seamount.

For legend see figure 20

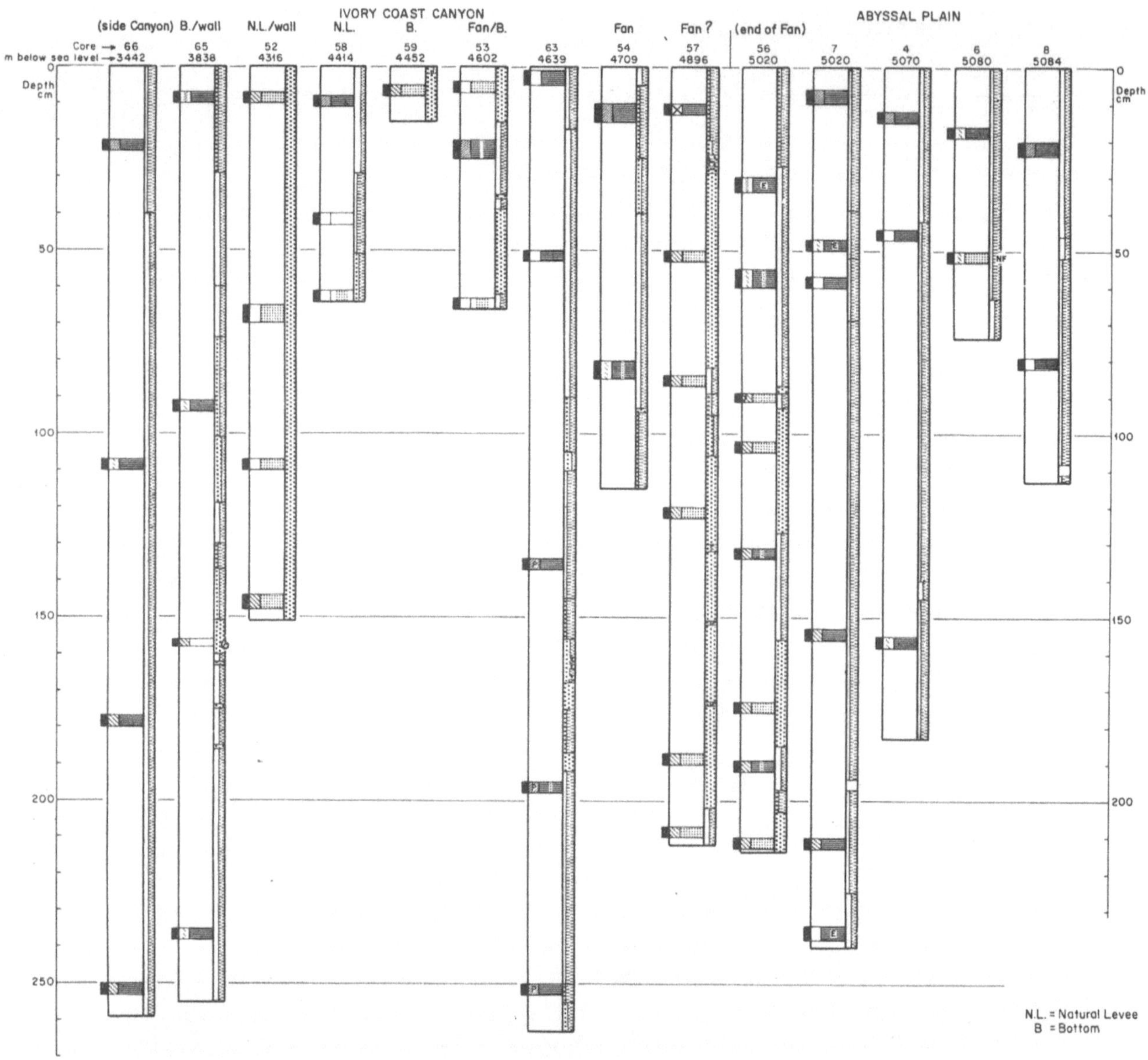

Fig. 20
Ivory Coast canyon and Abyssal plain.

vertical scale 1 : 10, horizontal scale: none

15 number of core station

1.441 sea depth in metres

- interval sampled
- RECENT/HOLOCENE (tropical planktonic fauna)
- idem, but not certain
- PLEISTOCENE (colder-water planktonic fauna)
- idem, but not certain
- PLEISTOCENE (Interglacial with tropical planktonic fauna)
- idem, but not certain
- mixed planktonic fauna
- ABYSSAL Planulina wuellerstorfi fauna
- ABYSSAL Planulina wuellerstorfi fauna (Epistominella exigua faunule)
- DEEPER BATHYAL - SHALLOW ABYSSAL Gavelinopsis translucens fauna et al.

- sandy residue with >50% shallow water benthonic foraminifera
- sandy residue with 10-50% shallow water benthonic foraminifera
- sandy residue with <10% shallow water benthonic foraminifera
- sand, silt, mud } after description of lithology by F. JONKER
- NF residue almost barren
- GC residue rich in glauconite
- residue rich in molluscs
- residue mainly consisting of plant remains
- residue rich in pellets
- Globigerina ooze

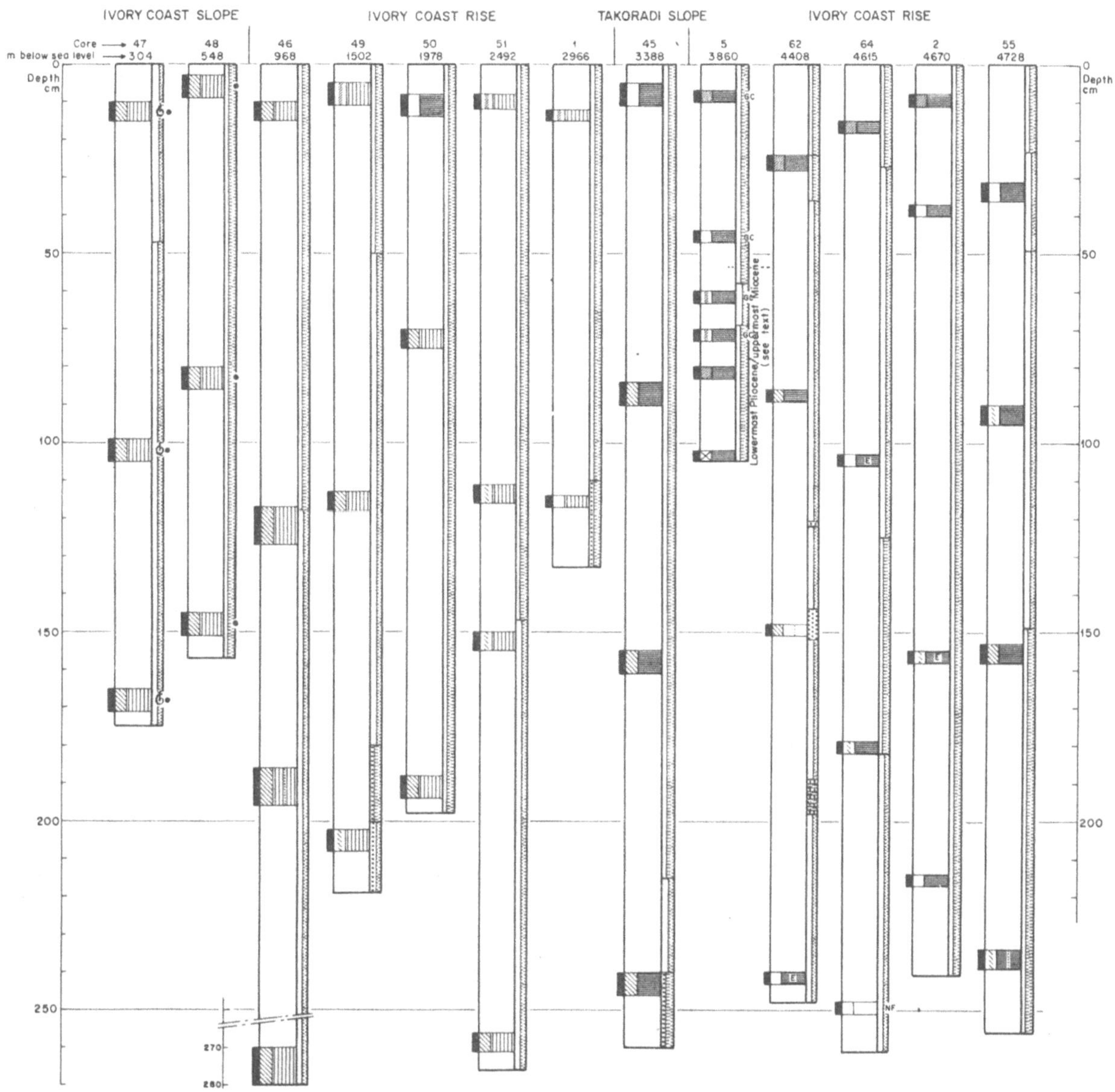

Fig. 21
Ivory Coast slope, Ivory Coast rise and Takoradi slope.

For legend see figure 20

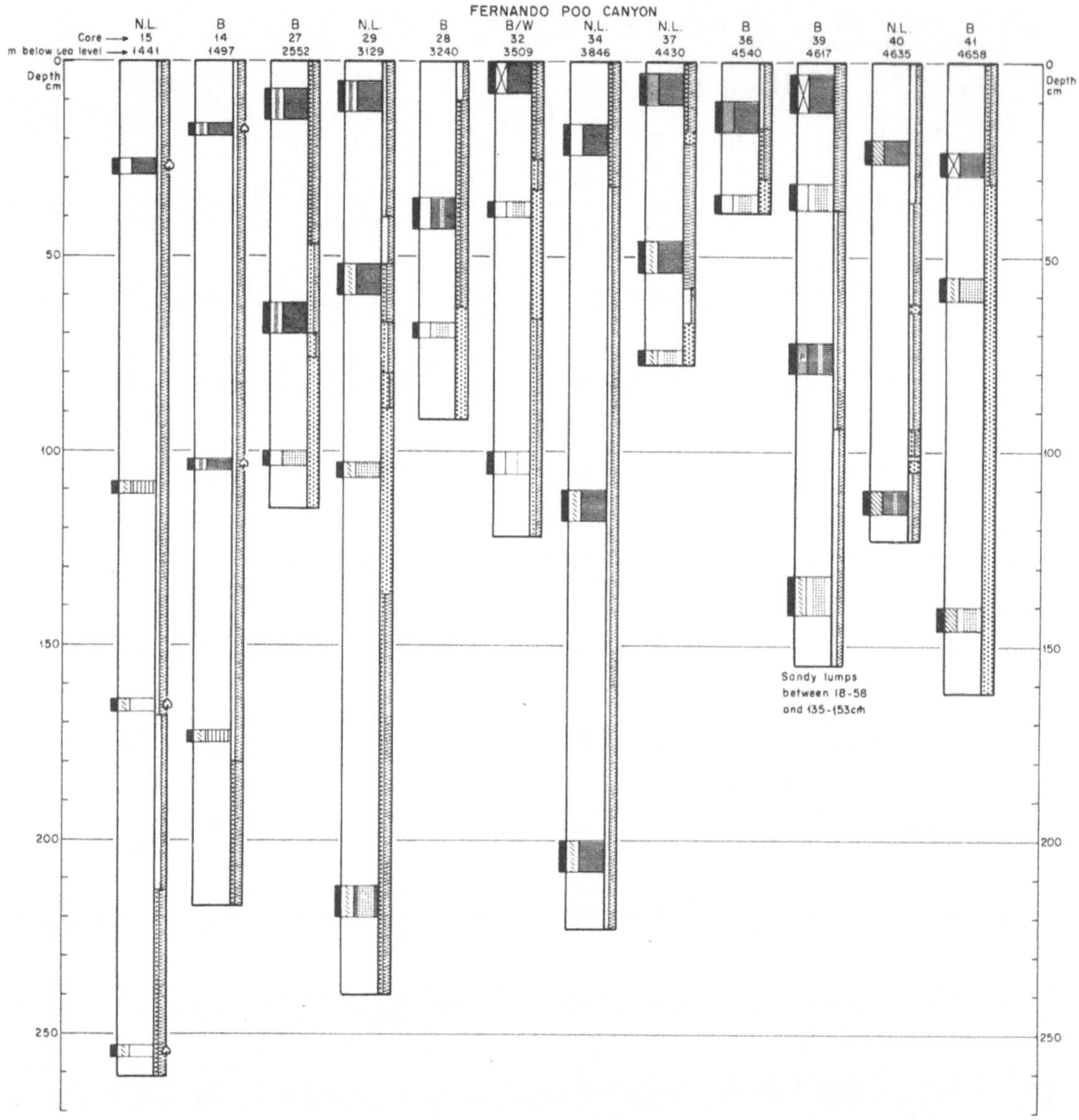

Fig. 22
Fernando Poo canyon

For legend see figure 20

TABLES

NELSON core	47			48			46				49		
	IVORY COAST CONTINENTAL SLOPE						IVORY COAST CONTINENTAL RISE						
sample interval (in cm from top)	10-15	99-105	165-171	3-9	80-86	145-151	10-15	117-127	186-196	270-280	5-11	113-118	202-208
water depth at sample station (metres)	304			543			969				1502		
dry weight of sample (grammes)	80	118	120	43	60	57	50	54	53	31	48	43	68
weight-percentage of residue	74	78	26	25	12	9	8	18	18	12	1	7	8
PLANKTONIC FORAMINIFERIDA													
Candeina nitida D'ORBIGNY													
Globigerina bulloides D'ORBIGNY	4	6		18	10	10	5	6	6	16		7	5
Globigerina inflata D'ORBIGNY	3	10	4	7	2		1	7	5	6	6		+
Globigerina pachyderma (EHRENBERG)	6	3	15	2	3		4	6	14	6		6	2
Globigerinella aequilateralis (BRADY)													+
Globigerinita glutinata (EGGER)				8	1		4		2	3		4	1
Globigerinoides conglobatus (BRADY)	+	1	+				1	+	+	+		+	+
Globigerinoides extremus BOLLI & BERMUDEZ													
Globigerinoides immaturus LEROY													
Globigerinoides ruber (D'ORBIGNY) (pink)	9	3	5	13	13	18	14	8	9	4	25	12	14
Globigerinoides sacculiferus (BRADY)	+	1	+	1	1			+	+	1		1	1
Globigerinoides trilobus (REUSS)	41	34	38	32	41	51	37	27	35	34	48	54	43
Globoquadrina altispira (CUSHMAN & JARVIS)													
Globoquadrina dutertrei (D'ORBIGNY)	1	11	9	2	5	6	16	11	2	11		8	9
Globorotalia acostaensis													
Globorotalia menardii (D'ORBIGNY)	2	+	4	+	+	3	7				7	+	
Globorotalia punctulata D'ORBIGNY	3	7	1	6	4		4		4	+	6	4	1
Globorotalia scitula (BRADY)		1						1	1			1	+
Globorotalia truncatulinoides (D'ORBIGNY)													
Orbulina bilobata (D'ORBIGNY)													
Orbulina universa D'ORBIGNY	2	1	+		+			+	+	1		1	1
Pulleniatina obliquiloculata (PARKER & JONES)	+	+	+	+	1	1			+			+	+
Pulleniatina primalis BANNER & BLOW													
Sphaeroidinella dehiscens (PARKER & JONES)				+	+								
Sphaeroidinella subdehiscens BLOW													
BENTHONIC FORAMINIFERIDA (agglutinated) (%)	+		+	+	1	3	+	3	2	1	2	4	3
Adercotryma glomerata (BRADY)													
Cribrostomoides subglobosum (SARS)													
Cyclammina trullisata (BRADY)													
Cystammina galeata (BRADY)													
Eggerella bradyi (CUSHMAN)					1	1		1				1	1
Gaudryina atlantica (BAILEY)							+	+	+	+		+	+
Glomospira charoides (JONES & PARKER)													
Hormosinidae													
Hyperammina elongata BRADY													
Karreriella apicularis (CUSHMAN)													
Martinottiella communis (D'ORBIGNY)	+			+		1							
Recurvoides turbinatus (BRADY)													
Rhabdammina abyssorum SARS													
Saccammina sphaerica BRADY													
Sigmoilopsis schlumbergeri (SILVESTRI)				+	+	1	+	2	2	1	2	2	2
Siphotextularia rolshauseni PHLEGER & PARKER													
Siphotextularia spec. indet.												1	
Textulariidae			+										
Trochammina globigeriniformis PARKER & JONES													
BENTHONIC FORAMINIFERIDA (calcareous) (%)	92	91	98	99	98	97	100	97	97	99	98	96	97
Amphicoryna scalaris (BATSCH)	+		+										
Anomalina globulosa CHAPMAN & PARR												2	1
Anomalinoides io (CUSHMAN)							+	2	3	3			+
Astrononion stelligerum (D'ORBIGNY)	1												
Bolivina albatrossi CUSHMAN							+	+	3	2			
Bolivina goësi CUSHMAN	2		11			+							
Bolivina pseudoplicata HERON-ALLEN & EARLAND													
Bolivinita quadrilatera (SCHWAGER)						1		+				1	
Brizalina alata (SEGUENZA)													
Brizalina pettersoni (PARKER)													
Brizalina subaenariensis (CUSHMAN)		4	4										
Bulimina aculeata D'ORBIGNY			1	1	3	3	+		2	4		4	16
Bulimina alazanensis CUSHMAN			1		1	1							
Bulimina exilis BRADY													
Bulimina inflata SEGUENZA	5	4					22	9	6	7	2	8	8
Bulimina marginata D'ORBIGNY	+	1	2										
Cassidella bradyi (CUSHMAN)													
Cassidella mexicana (CUSHMAN)									2				
Cassidella spec. indet.													
Cassidulina carinata SILVESTRI	6	12	14	7	5	2		8	13	9	1	4	4
Cassidulinoides bradyi (NORMAN)													
Cassidulinoides mexicana (CUSHMAN)													
Ceratobulimina pacifica CUSHMAN & HARRIS								1	1	1		7	1
Chilostomella oolina SCHWAGER		+	1							1			
Cibicides ex gr. lobatulus (WALKER & JACOB)										1			
Cibicidoides robertsonianus (BRADY)					1		1	3	3	3	1	2	
Ehrenbergina trigona GOËS													
Ehrenbergina undulata PARKER	5	2	2		1	2							
Epistominella exigua (BRADY)											+		
Epistominella smithi (STEWART & STEWART)				+				+		2			1
Eponides polius PHLEGER & PARKER			1						2	1	3	2	1
Eponides pusillus PARR													
Eponides regularis PHLEGER & PARKER				1			1						
Euuvigerina peregrina (CUSHMAN)	24	22	18	3	6	6	22	8	10	16	50	10	10
Francesita advena (CUSHMAN)													
Fursenkoina complanata (EGGER)									1				

NELSON core	50			51				1		45			
	IVORY COAST CONTINENTAL RISE												
sample interval (in cm from top)	8-14	70-75	188-194	8-13	111-116	150-155	256-260	12-15	114-117	5-11	84-90	155-161	240-246
water depth at sample station (metres)	1973			2492				2966		3388			
dry weight of sample (grammes)	47	45	48	37	41	43	39	31	30	44	52	47	61
weight-percentage of residue	2	1	8	2	5	2	2	1	2	1	2	3	2
PLANKTONIC FORAMINIFERIDA													
Candeina nitida D'ORBIGNY													
Globigerina bulloides D'ORBIGNY	6	3			5		5					4	5
Globigerina inflata D'ORBIGNY	1		2				4			6	5		2
Globigerina pachyderma (EHRENBERG)		7	10			4	6				14	9	24
Globigerinella aequilateralis (BRADY)													
Globigerinita glutinata (EGGER)	3	4	2		2	1	2			1		6	
Globigerinoides conglobatus (BRADY)		+	+		+		1	+	+				
Globigerinoides extremus BOLLI & BERMUDEZ													
Globigerinoides immaturus LEROY													
Globigerinoides ruber (D'ORBIGNY) (pink)	13	10	17	17	17	16	12	10	18	11	10	15	14
Globigerinoides sacculiferus (BRADY)			5			1		4	2	1		3	
Globigerinoides trilobus (REUSS)	50	43	30	59	50	66	35	27	22	51	25	31	22
Globoquadrina altispira (CUSHMAN & JARVIS)													
Globoquadrina dutertrei (D'ORBIGNY)	3	8	12	3	11	4	15	14	32	4	12	12	8
Globorotalia acostaensis													
Globorotalia menardii (D'ORBIGNY)	3	2		5	+			5		4	1		
Globorotalia punctulata D'ORBIGNY	1	3	2	10	1	4	5	1	4	12	6	3	
Globorotalia scitula (BRADY)	2	3	1	2			4			5	1	2	2
Globorotalia truncatulinoides (D'ORBIGNY)		1				1					+		
Orbulina bilobata (D'ORBIGNY)													
Orbulina universa D'ORBIGNY			1		+	1	+	2	+		+	+	+
Pulleniatina obliquiloculata (PARKER & JONES)				1				2			+		
Pulleniatina primalis BANNER & BLOW													
Sphaeroidinella dehiscens (PARKER & JONES)	+	+		+	+			1	+		+		
Sphaeroidinella subdehiscens BLOW													
BENTHONIC FORAMINIFERIDA (agglutinated) (%)	1	2	+		1	2	2	16	1		2	2	+
Adercotryma glomerata (BRADY)													
Cribrostomoides subglobosum (SARS)													
Cyclammina trullisata (BRADY)													
Cystammina galeata (BRADY)													
Eggerella bradyi (CUSHMAN)								2				2	
Gaudryina atlantica (BAILEY)													
Glomospira charoides (JONES & PARKER)													
Hormosinidae													
Hyperammina elongata BRADY													
Karreriella apicularis (CUSHMAN)													
Martinottiella communis (D'ORBIGNY)							1	14					
Recurvoides turbinatus (BRADY)													
Rhabdammina abyssorum SARS													
Saccammina sphaerica BRADY													
Sigmoilopsis schlumbergeri (SILVESTRI)	1	2			1	2	1	+	1		2		
Siphotextularia rolshauseni PHLEGER & PARKER													
Siphotextularia spec. indet.													
Textulariidae			+										
Trochammina globigeriniformis PARKER & JONES													
BENTHONIC FORAMINIFERIDA (calcareous) (%)	99	98	100	100	99	98	96	84	99	100	97	98	100
Amphicoryna scalaris (BATSCH)													
Anomalina globulosa CHAPMAN & PARR	+		+										+
Anomalinoides io (CUSHMAN)	5	4											
Astrononion stelligerum (D'ORBIGNY)						3	+					5	3
Bolivina albatrossi CUSHMAN													
Bolivina goësi CUSHMAN									+				
Bolivina pseudoplicata HERON-ALLEN & EARLAND													
Bolivinita quadrilatera (SCHWAGER)													
Brizalina alata (SEGUENZA)													
Brizalina pettersoni (PARKER)													
Brizalina subaenariensis (CUSHMAN)													
Bulimina aculeata D'ORBIGNY			16			1	5					2	
Bulimina alazanensis CUSHMAN	4	6	18	5	6	11		1		2	1		
Bulimina exilis BRADY													
Bulimina inflata SEGUENZA	6	3	2	1	1	8	+						
Bulimina marginata D'ORBIGNY										1	1	2	
Cassidella bradyi (CUSHMAN)			+			+							3
Cassidella mexicana (CUSHMAN)													
Cassidella spec. indet.													
Cassidulina carinata SILVESTRI	9	9	16	4	18	2	8		1		4	5	
Cassidulinoides bradyi (NORMAN)													
Cassidulinoides mexicana (CUSHMAN)													
Ceratobulimina pacifica CUSHMAN & HARRIS	1	+											
Chilostomella oolina SCHWAGER			+		1				2				
Cibicides ex gr. lobatulus (WALKER & JACOB)			+										
Cibicidoides robertsonianus (BRADY)	1	+		5	4	+		2	1		1	8	
Ehrenbergina trigona GOËS													
Ehrenbergina undulata PARKER													
Epistominella exigua (BRADY)	5	1	+			10	+						3
Epistominella smithi (STEWART & STEWART)											2		
Eponides polius PHLEGER & PARKER	3	8		1	1					3	2	+	+
Eponides pusillus PARR													
Eponides regularis PHLEGER & PARKER					+	+							
Euuvigerina peregrina (CUSHMAN)	19	17	6	39	16	26	42	52	54	20	12	17	36
Francesita advena (CUSHMAN)													
Fursenkoina complanata (EGGER)													

NELSON core	5						62				2				55			
	IVORY COAST CONTINENTAL RISE																	
sample interval (in cm from top)	7-10	44-47	60-63	70-73	80-83	102-105	24-28	86-89	148-151	240-243	8-11	37-40	155-158	214-217	31-36	90-95	133-158	234-239
water depth at sample station (metres)	3960						4408				4670				4728			
dry weight of sample (grammes)	30	33	34	46	39	53	47	28	49	21	32	17	22	22	45	32	40	42
weight-percentage of residue	15	23	2	2	4	4	2	4	41	+	7	1	4	+	1	+	+	1
PLANKTONIC FORAMINIFERIDA																		
Candeina nitida D'ORBIGNY																		
Globigerina bulloides D'ORBIGNY				7		20		7	30									
Globigerina inflata D'ORBIGNY	1						2									1	1	4
Globigerina pachyderma (EHRENBERG)							2	10	27				1		4	6	12	3
Globigerinella aequilateralis (BRADY)														X				1
Globigerinita glutinata (EGGER)								1	1	X							1	
Globigerinoides conglobatus (BRADY)	+			+		+		+			1		+					
Globigerinoides extremus BOLLI & BERMUDEZ																		
Globigerinoides immaturus LEROY																		
Globigerinoides ruber (D'ORBIGNY) (pink)	1	X						7	9			11	10	X		8	2	
Globigerinoides sacculiferus (BRADY)	1					1					3		8			4		2
Globigerinoides trilobus (REUSS)	2			28	31	16	3	54	23	X	13	56	18	X		35	21	33
Globoquadrina altispira (CUSHMAN & JARVIS)				2		2												
Globoquadrina dutertrei (D'ORBIGNY)	32			32	18		27	5			37	14	33	X	67	30	40	27
Globorotalia acostaensis																		
Globorotalia menardii (D'ORBIGNY)	13		X	8	17	10	57				17				2			
Globorotalia punctulata D'ORBIGNY	2				3	1		2	2		10		1		12	9	12	13
Globorotalia scitula (BRADY)						2		2	2			1				1		
Globorotalia truncatulinoides (D'ORBIGNY)																		
Orbulina bilobata (D'ORBIGNY)																		
Orbulina universa D'ORBIGNY	+			+		+							+					1
Pulleniatina obliquiloculata (PARKER & JONES)	1						2	+			+	4	5		4			2
Pulleniatina primalis BANNER & BLOW						2												
Sphaeroidinella dehiscens (PARKER & JONES)	1										3	1						
Sphaeroidinella subdehiscens BLOW			X	2	3	+												
BENTHONIC FORAMINIFERIDA (agglutinated) (%)	1		+	5	2	9		1			8		5		+			
Adercotryma glomerata (BRADY)																		
Cribrostomoides subglobosum (SARS)																		
Cyclammina trullisata (BRADY)											3							
Cystammina galeata (BRADY)											2							
Eggerella bradyi (CUSHMAN)	1			3	2	5		1			+		4					
Gaudryina atlantica (BAILEY)																		
Glomospira charoides (JONES & PARKER)																		
Hormosinidae																		
Hyperammina elongata BRADY																		
Karreriella apicularis (CUSHMAN)											1							
Martinottiella communis (D'ORBIGNY)				+	+	1												
Recurvoides turbinatus (BRADY)																		
Rhabdammina abyssorum SARS											2							
Saccammina sphaerica BRADY																		
Sigmoilopsis schlumbergeri (SILVESTRI)			+	2		2												
Siphotextularia rolshauseni PHLEGER & PARKER																		
Siphotextularia spec. indet.													1					
Textulariidae						1												
Trochammina globigeriniformis PARKER & JONES											+							
BENTHONIC FORAMINIFERIDA (calcareous) (%)	99	100	100	95	98	91	100	99	96	100	92	100	95	100	100	100	100	100
Amphicoryna scalaris (BATSCH)					4													
Anomalina globulosa CHAPMAN & PARR						+												
Anomalinoides io (CUSHMAN)			2	+	1								3					
Astrononion stelligerum (D'ORBIGNY)				2		3	1	1				1						
Bolivina albatrossi CUSHMAN																		
Bolivina goësi CUSHMAN							1	1	19						+			
Bolivina pseudoplicata HERON-ALLEN & EARLAND												1						
Bolivinita quadrilatera (SCHWAGER)																		
Brizalina alata (SEGUENZA)																		
Brizalina pettersoni (PARKER)									19									
Brizalina subaenariensis (CUSHMAN)						1			8									
Bulimina aculeata D'ORBIGNY									1									
Bulimina alazanensis CUSHMAN				1		1	1											
Bulimina exilis BRADY									4									
Bulimina inflata SEGUENZA			+	1		1	1		2									
Bulimina marginata D'ORBIGNY																		
Cassidella bradyi (CUSHMAN)				1									5					2
Cassidella mexicana (CUSHMAN)									2									
Cassidella spec. indet.																		
Cassidulina carinata SILVESTRI								2	14									
Cassidulinoides bradyi (NORMAN)																		
Cassidulinoides mexicana (CUSHMAN)				1														
Ceratobulimina pacifica CUSHMAN & HARRIS																		
Chilostomella oolina SCHWAGER									1									
Cibicides ex gr. lobatulus (WALKER & JACOB)						+											4	
Cibicidoides robertsonianus (BRADY)	23		3	2	6	5	3	8			8				8	7		5
Ehrenbergina trigona GOËS							1											
Ehrenbergina undulata PARKER			1	1		1												
Epistominella exigua (BRADY)	4	24				1	24	16		56	16	20	4	21	12		11	20
Epistominella smithi (STEWART & STEWART)									14									
Eponides polius PHLEGER & PARKER															4			
Eponides pusillus PARR										29		24		39				
Eponides regularis PHLEGER & PARKER							1											
Euuvigerina peregrina (CUSHMAN)				2				1	3									
Francesita advena (CUSHMAN)																		
Fursenkoina complanata (EGGER)														1				

Ophthalmidium pusillum (EARLAND)																																												
Oridorsalis umbonatus (REUSS)	+	+					1	2	2	3	3	4	5	5	4	+	2	2	1	3	3	2	10	9	6	+	25	1	20	10	13	10	13	27	1		8	4	33	4	10		22	26
Osangularia culter (PARKER & JONES)							12	3	1	3		6	8	2		2		3	1	1			1								3													
Planulina ariminensis D'ORBIGNY	2	2	1	+		1																		1																				
Planulina wuellerstorfi (SCHWAGER)								1				1	+	2	3	2	11	4	4	13	7	17	14	25	13	9	9	1	5	8	7	5	17	10	1	1	15	3	13	13	5	81	34	11
Pleurostomella alternans SCHWAGER																														1		+												
Polymorphinidae																															1													
Praeglobobulimina pupoides (D'ORBIGNY)			+		+																1			1		1						+												
Pseudonodosaria torrida (CUSHMAN)						+																																						
Pullenia bulloides (D'ORBIGNY)							1	1		2	1	+	1	2	3	1	2		1	2	3	1	2	2		3	3		1	5	2	5		2			2	6	6		5		1	9
Pullenia quinqueloba (REUSS)																	1	1	1					1			3		+	1	2	2						6		4			3	6
Pyrgo lucernula (SCHWAGER)												+	+																			+												
Pyrgo murrhyna (SCHWAGER)								+		1			+		1	+	1	2	+		2	1	6	3	4	1	5			1	1	1		3			2		9					3
Pyrgo serrata (BAILEY)																																												
Pyrgo spp. indet.		+			+			+	+	+		1												+						+				+				+					3	
Pyrgoella sphaera (D'ORBIGNY)																																												
Pyrulina cylindroides (ROEMER)																																												
Quinqueloculina spp. indet.		3	1					2	2	+		+	+	3	4	+		5	1		+		2	6	4	+						1		4			2	2	4					
Rotamorphina laevigata (PHLEGER & PARKER)													+			1			+																									
Robertina bradyi CUSHMAN & PARKER												+			+																													
Saracenaria italica DEFRANCE					+	+																																						
Siphonodosaria spp. indet.																														6	+	3												
Siphouvigerina auberiana (D'ORBIGNY)				63	74	72	3	2	3	4	11	15	21	16	5	18	6	5	14	3			5	1	+	19		53	36	21	11	16		1	2									
Sphaeroidina bulloides D'ORBIGNY				+	+		7	2	3	8	2	4	1	1	2	+		2	+			1		1	3	1	2		1	2	5	1												
Spirillina decorata BRADY																																												
Spiroloculina spp. indet.	+	+						+		+																															+			
Spirosigmoilina tenuis (CZJZEK)	+								1																																			
Trifarina angulosa (WILLIAMSON)	25	16	22	1	3	1			1	+							1																											
Triloculina tricarinata D'ORBIGNY																																												
Vaginulina spinigera BRADY																																												
Valvulineria arctica GREEN														+	+																													
Valvulineria complanata (D'ORBIGNY)																																												
SHALLOW WATER BENTHONIC FORAMINIFERIDA (%)	8	9	2	1	1	+			1								0	0	0	0	0	0	0	1	0	0	0	0	0	0	+				1									+
Ammonia beccarii (LINNE)		+	1																												+													
Amphistegina lessonii D'ORBIGNY	+																																											
Asterigerinata mamilla (WILLIAMSON)																																												
Cancris oblongus (WILLIAMSON)	+	+	+	+	+	+			1																																			
Cibicides ex gr. lobatulus (WALKER & JACOB)																																												
Cribroelphidium gunteri (COLE)																																												
Cribrononion advenum (CUSHMAN)																																												+
Cribrononion incertum (WILLIAMSON)																																												
Dyocibicides spec. indet.																																			+									
Elphidiidae																																												
Elphidium crispum (LINNE)																																												
Eponides repandus (FICHTEL & MOLL)																																												
Florilus asterizans (FICHTEL & MOLL)																																												
Florilus atlanticus (CUSHMAN)		2	+																					1											1									
Hanzawaia ex gr. concentrica (CUSHMAN)	5	5	1	1	1																																							
Miliolidae	+																																											
Neoconorbina orbicularis (TERQUEM)																																			2									
Nonionella spec. indet.																																												
Pararotalia armata (D'ORBIGNY)																																												
Planorbulina mediterranensis D'ORBIGNY		+																																										
Poritextularia panamensis (CUSHMAN)																																												
Ptychomiliola separans (BRADY)																																												
Quinqueloculina spp. indet.	+																																											
Reussella miocenica CUSHMAN																																			1									
Rosalina globularis D'ORBIGNY																																			+									
Textulariidae	3	2																																										
Number of Radiolaria per gramme sediment				210	110	18	78	5	50		52			78	89	31	260	70	9	100	30	2	3	5	5	11	240	1							160	10	+	3	280	17		27	1	1
Echinoidea (spines)	463	1740	85	47			18	49	10	2	6	15	66	19	5	32	5	33	5	20	+		1	24	5	34	+	3	2	4	11	10	1	9	3300		+				1	+		1
Echinoidea (plates)	1		3		+			24	+																																			
Mollusca	2300	8700	1000	47				320	230	34		120	+		+	49		75					+		+	+								2	2400									
Ostracoda	600	280	50	50		9		14	10	23		8	3		+	1		+			10	1	+	+	+						+			+	70		+		+				+	
Pisces (teeth)											+						+															+		+		+								
Pisces (otoliths)	1	70	17					11	+	+		+	+	+		1	+	+	+	+		+			+				+															+
Bryozoa																																												

Table 1

	ABIDJAN CANYON																									
NELSON core	66				65				52				58			59	53			63					54	
sample interval (in cm from top)	20-23	107-110	177-180	250-255	7-10	91-94	156-158	235-238	7-10	65-70	107-110	144-148	8-11	40-43	61-64	5-8	4-7	20-25	63-66	4-5	50-53	134-137	195-198	250-253	10-15	80-85
water depth at sample station (metres)	3442				3839				4316				4414			4452	4602			4639					4709	
dry weight of sample (grammes)	19	25	21	24	20	32	30	27	66	108	78	79	19	29	35	76	51	53	31	22	22	28	30	31	42	33
weight-percentage of residue	2	3	2	+	1	6	67	+	11	37	15	36	1	1	25	32	11	6	16	1	1	10	3	18	16	2
PLANKTONIC FORAMINIFERIDA																										
Candeina nitida D'ORBIGNY					2	3																				
Globigerina bulloides D'ORBIGNY				3	5	3	7		8		4	12	5			15	3									
Globigerina inflata D'ORBIGNY			13	15			6					4		1		5	3	7	6	3	4	2		+	1	2
Globigerina pachyderma (EHRENBERG)		3	17	5				5	5		4	6		1	2	5	2		3		2	8				5
Globigerinella aequilateralis (BRADY)							+											1						1		
Globigerinita glutinata (EGGER)		7		3	2	4					2				1		3	1	3					3		2
Globigerinoides conglobatus (BRADY)	+			2	+	+	+			+	1					+			+		+	+		2	+	
Globigerinoides extremus BOLLI & BERMUDEZ																										
Globigerinoides immaturus LEROY																										
Globigerinoides ruber (D'ORBIGNY) (pink)	21	26	19	13	12	3	10	14	12	19	7	10	5		14	3	9	2	10	7		2	2	5		8
Globigerinoides sacculiferus (BRADY)	18	26	15	10	3	1	5	8	2	3	3	+			1	2	+	1		1	2	4		+		
Globigerinoides trilobus (REUSS)	35	16	8	22	37	48	19	46	39	34	44	48	34	2	63	45	33	39	35	50	7	26	34	10	11	18
Globoquadrina altispira (CUSHMAN & JARVIS)																										
Globoquadrina dutertrei (D'ORBIGNY)	12	8	19	15	3	11	23	7	3	14	5	5	6	35	10	9	6	13	21	6	32	7	2	23	33	39
Globorotalia acostaensis																										
Globorotalia menardii (D'ORBIGNY)	11			2	8	+	3		4	4	5	3	38	44	+	3	4	29	11	14	33	45	40	30	35	
Globorotalia punctulata D'ORBIGNY		3	4	3	9	6	14	8		+	5	1	5	4	2	+	2		1	4		1		5	6	8
Globorotalia scitula (BRADY)		2			1	1	1			3	1						1									
Globorotalia truncatulinoides (D'ORBIGNY)												1					1					1				
Orbulina bilobata (D'ORBIGNY)																										
Orbulina universa D'ORBIGNY		1				1	1	1		+	+	+	+	+		+	3	1	+						1	
Pulleniatina obliquiloculata (PARKER & JONES)	+				+	1				+	1		5	4	1	+		2		4	2	3			3	
Pulleniatina primalis BANNER & BLOW																										
Sphaeroidinella dehiscens (PARKER & JONES)	+				3	+			+	+	+	+	+	1	+	2	+	1	4				1	1	3	
Sphaeroidinella subdehiscens BLOW																										
BENTHONIC FORAMINIFERIDA (agglutinated) (%)	1	0	0	0	3	2	1	0	0	0	0	0	8	0	2	0	0	4	0	0	1	2	0	2	4	0
Adercotryma glomerata (BRADY)																										
Cribrostomoides subglobosum (SARS)																										
Cyclammina trullisata (BRADY)													1													
Cystammina galeata (BRADY)																										
Eggerella bradyi (CUSHMAN)	1				1	1								+	1		+	4				2		2	3	
Gaudryina atlantica (BAILEY)																										
Glomospira charoides (JONES & PARKER)																										
Hormosinidae																										
Hyperammina elongata BRADY																										
Karreriella apicularis (CUSHMAN)					1																					
Martinottiella communis (D'ORBIGNY)																										
Recurvoides turbinatus (BRADY)																										
Rhabdammina abyssorum SARS													1													
Saccammina sphaerica BRADY																										
Sigmoilopsis schlumbergeri (SILVESTRI)					1		1				+				1											
Siphotextularia rolshauseni PHLEGER & PARKER																					1					
Siphotextularia spec. indet.																										
Textulariidae																										
Trochammina globigeriniformis PARKER & JONES													6													
BENTHONIC FORAMINIFERIDA (calcareous) (%)	99	100	100	100	97	98	92	100	32	28	29	29	92	99	67	29	40	91	64	100	99	98	97	98	96	94
Amphicoryna scalaris (BATSCH)																										
Anomalina globulosa CHAPMAN & PARR						2																		1		
Anomalinoides io (CUSHMAN)					7	4																10		9		
Astrononion stelligerum (D'ORBIGNY)					8	6		+					4	1	1	1	1	2			2	3	1	1	1	3
Bolivina albatrossi CUSHMAN									+	+	+															
Bolivina goesi CUSHMAN		+				+	22		2	6	10	4		+	24	5	8	2	11		1		5		+	
Bolivina pseudoplicata HERON-ALLEN & EARLAND																										
Bolivinita quadrilatera (SCHWAGER)						1																				
Brizalina alata (SEGUENZA)																			1							
Brizalina pettersoni (PARKER)						9	3		9	4	3	3		1	4	4	11	4	10				8		1	
Brizalina subaenariensis (CUSHMAN)						1	3		4	4	3	2			2	1	1		3				1			
Bulimina aculeata D'ORBIGNY							2			1					1		1									
Bulimina alazanensis CUSHMAN						2			1	1		1														
Bulimina exilis BRADY							3		2		1				1	1	1	+	1				2			
Bulimina inflata SEGUENZA							1				+	+		+			1						1			
Bulimina marginata D'ORBIGNY						+	1					1				2		1	2				1			
Cassidella bradyi (CUSHMAN)							1								1										1	3
Cassidella mexicana (CUSHMAN)						+										1										
Cassidella spec. indet.									+	2	2	2				+	2	+	2							
Cassidulina carinata SILVESTRI						2	9		6	5	5	6		+	8	2	2	1	4				4			
Cassidulinoides bradyi (NORMAN)																										
Cassidulinoides mexicana (CUSHMAN)											1			+		1										
Ceratobulimina pacifica CUSHMAN & HARRIS																										
Chilostomella oolina SCHWAGER							2		1																	
Cibicides ex gr. lobatulus (WALKER & JACOB)																										
Cibicidoides robertsonianus (BRADY)	14				1	3							5	5	2			7			7			6	3	
Ehrenbergina trigona GOËS																										
Ehrenbergina undulata PARKER																										
Epistominella exigua (BRADY)	4	19		4	4	2		7					24	29	1			19	3	33	19	21	15	15	19	11
Epistominella smithi (STEWART & STEWART)						5	6		2	2	2	3			8	1	7	1	6				1			
Eponides polius PHLEGER & PARKER					+																					
Eponides pusillus PARR																				•						
Eponides regularis PHLEGER & PARKER												+													1	
Euuvigerina peregrina (CUSHMAN)	5			1	12	3	11	2	1	+	2	3		+	1	1			5							
Francesita advena (CUSHMAN)																										
Fursenkoina complanata (EGGER)																										
Gavelinopsis translucens (PHLEGER & PARKER)																										
Globobulimina pacifica CUSHMAN		+	1		4		4		+	+					1	+	+	1								
Globocassidulina oblonga (REUSS)		1						1									1									
Globocassidulina subglobosa (BRADY)				1		1			+							1						1			+	
Gyroidina lamarckiana (D'ORBIGNY)														2					1	5	4	1		6	3	
Gyroidina soldanii D'ORBIGNY	3	+			1	2		+	+				4	2	1				1	5	2		1	4		
Heterolepa pseudoungeriana (CUSHMAN)							7					+		2	2	3			3							
Hoeglundina elegans (D'ORBIGNY)	5					1	1		+	+		+				+										
Hyalinea balthica (SCHROETER)									2	1	+				1	1							1			
Lagena s.l.	2	2			5		1	+					3	1				2	1		6	2	3	4	2	
Laticarinina pauperata (PARKER & JONES)																										
Lenticulina peregrina (SCHWAGER)																										
Lenticulina s.l.		+			2		1			+		+		1	1		1	1					1			
Lingulina seminuda HANTKEN																										
Melonis affinis (REUSS)	8	2			7	1	2				+		7	1	1			8		5	3		1	9	3	5
Melonis pompilioides (WILLIAMSON)	13	3	2	3	3	3		2					3	2				1				3				
Miliolidae					3	3	+													2		1				
Nodosaria longiscata D'ORBIGNY																										
Nodosaria (incl. Dentalina)																								1		
Nummoloculina irregularis (D'ORBIGNY)																										
Nuttallides umboniferus (CUSHMAN)	1												8	18	1			4	5	7	24			12	32	
Ophthalmidium pusillum (EARLAND)																										
Oridorsalis umbonatus (REUSS)	19	14		5	5	9	1	1		1			11	10	+			7	1	7	6	24	34	6	7	10
Osangularia culter (PARKER & JONES)															1	1										
Planulina ariminensis D'ORBIGNY							2																			
Planulina wuellerstorfi (SCHWAGER)	19	18	13	82	13	17	3	15		+	+		14	15	2			12	2	10	13	11	10	14	7	50
Pleurostomella alternans SCHWAGER																										
Polymorphinidae					1																					
Praeglobobulimina pupoides (D'ORBIGNY)							+																			
Pseudonodosaria torrida (CUSHMAN)																										
Pullenia bulloides (D'ORBIGNY)		1	2	4	1	2		1	1				8	4	1			12		12	7	3	1			1
Pullenia quinqueloba (REUSS)	2	1			3	+																	1	4	9	
Pyrgo lucernula (SCHWAGER)																										
Pyrgo murrhyna (SCHWAGER)		+			1	4	+	2			+			1		+						1	1	4	2	
Pyrgo serrata (BAILEY)																										
Pyrgo spp. indet.							+			+	+	+	+			1		2	1							
Pyrgoella sphaera (D'ORBIGNY)																										
Pyrulina cylindroides (ROEMER)																										
Quinqueloculina spp. indet.			1		12	+	2														2	2		1	1	
Rotamorphina laevigata (PHLEGER & PARKER)																										
Robertina bradyi CUSHMAN & PARKER															+											
Saracenaria italica DEFRANCE																										
Siphonodosaria spp. indet.									1																	
Siphouvigerina auberiana (D'ORBIGNY)		37	80				3	67		1						1	1	1	1			12	1			1
Sphaeroidina bulloides D'ORBIGNY	1				1	1	1			+															1	
Spirillina decorata BRADY																										
Spiroloculina spp. indet.												+				1										
Spirosigmoilina tenuis (CZJZEK)													1													
Trifarina angulosa (WILLIAMSON)																										
Triloculina tricarinata D'ORBIGNY																										
Vaginulina spinigera BRADY																										
Valvulineria arctica GREEN																										
Valvulineria complanata (D'ORBIGNY)																										
SHALLOW WATER BENTHONIC FORAMINIFERIDA (%)	0	0	0	0	0	0	7	0	68	72	71	71	0	1	31	71	60	5	38	0	0	0	3	0	0	6
Ammonia beccarii (LINNE)									12	6	12	7		1	6	6	10	1	2				1			
Amphistegina lessonii D'ORBIGNY									2	2	2	3				2	1		2					1		1
Asterigerinata mamilla (WILLIAMSON)											+															
Cancris oblongus (WILLIAMSON)							1		2	1	4	+				2	4		3				1			
Cibicides ex gr. lobatulus (WALKER & JACOB)									7	10	8	7			12	7	4									
Cribroelphidium gunteri (COLE)									2	1		3			1	2										
Cribrononion advenum (CUSHMAN)											2				1											
Cribrononion incertum (WILLIAMSON)																										
Dyocibicides spec. indet.																										
Elphidiidae												2				1	3	1	3							1
Elphidium crispum (LINNE)										+	+	+														
Eponides repandus (FICHTEL & MOLL)									+	4	2	1				2										
Florilus asterizans (FICHTEL & MOLL)																										
Florilus atlanticus (CUSHMAN)							1		12	13	7	13			4	9	11		5				1			
Hanzawaia ex gr. concentrica (CUSHMAN)									5	5	4	3			1	6	4	1	1							
Miliolidae											1						1									
Neoconorbina orbicularis (TERQUEM)							1		1							2										
Nonionella spec. indet.															1											
Pararotalia armata (D'ORBIGNY)							2		9	14	16	14			1	14	10	1	14							3
Planorbulina mediterranensis D'ORBIGNY									1		+	1			1											
Poritextularia panamensis (CUSHMAN)							+			+	1	+				1										
Ptychomiliola separans (BRADY)										+																
Quinqueloculina spp. indet.									9	11	8	10			1	9	8		5							1
Reussella miocenica CUSHMAN									1	1		+			1	2										
Rosalina globularis D'ORBIGNY							1		3	2	2	4				2										
Textulariidae									2	1	2	2			1	4	1	1	1							
Number of Radiolaria per gramme sediment	+	14	170	26	1	150		46					2						2				+		150	
Echinoidea (spines)	1	38	6	+	+	38	130	2	230	200	170	160	4	3	2	230	240	2	15		5	3	5	9	9	+
Echinoidea (plates)							2		+	+							340	+	+							
Mollusca	+	1			+	8	330	1	140	62	29	54	+		8	53	200	1	7							
Ostracoda					+		330		9	1	3	+			+	3	50	+	1			+		+		+
Pisces (teeth)				+																						
Pisces (otoliths)	+					+	3																			+
Bryozoa									4	3	4	5				6	•									

Table 2

GUINEA ABYSSAL PLAIN

NELSON core	64				42				57						56								7						4			6		8	
sample interval (in cm from top)	15-18	103-106	179-182	248-251	25-32	70-78	85-90	121-124	10-13	50-53	84-87	120-123	187-190	207-210	30-34	55-60	89-91	102-105	131-134	172-176	189-192	210-213	6-10	47-50	57-60	153-156	210-213	234-238	12-15	44-47	155-158	16-19	50-53	20-24	79-82
water depth at sample station (metres)	4615				4877				4896						5020								5020						5070			5082		5084	
dry weight of sample (grammes)	24	19	24	24	39	27	19	27	31	76	33	67	73	34	27	29	18	68	19	76	23	60	25	19	20	12	22	29	31	18	17	18	16	31	21
weight-percentage of residue	1	+	1	+	17	10	+	10	13	86	1	84	91	2	+	+	1	32	1	32	1	83	11	+	9	3	3	+	29	2	3	6	+	7	1
PLANKTONIC FORAMINIFERIDA																																			
Candeina nitida D'ORBIGNY																																			1
Globigerina bulloides D'ORBIGNY			2		2			10	3	13	22	13	9	2			3	17		9	5	11											X		
Globigerina inflata D'ORBIGNY	4					4	3	2	3	2		2	1	2			5	1	4	2	2					16	11				1	1			
Globigerina pachyderma (EHRENBERG)			5		19	7	9		11	13	14	5	10	15			7	11	8	9	8	10		X	2	+	3				1	3			
Globigerinella aequilateralis (BRADY)					+							2	1																						
Globigerinita glutinata (EGGER)					2	1		9				3															1								
Globigerinoides conglobatus (BRADY)	+								+	2		+	+	+				+	+			+			1				+		1				
Globigerinoides extremus BOLLI & BERMUDEZ																																			
Globigerinoides immaturus LEROY																																			
Globigerinoides ruber (D'ORBIGNY) (pink)			1		4	18	21	6		12	3	12	13	13			10	3	3	7	3	10		X	4	13	7	X	2	13	11		X		21
Globigerinoides sacculiferus (BRADY)					3	1	3		3	3		+	1	1				+	+	4		+			3		+		9	7	5	2		2	8
Globigerinoides trilobus (REUSS)	4		47		27	24	29	23	39	18	46	27	31	40			40	29	30	40	53	32	1		7	16	8	X	13	33	28	4		6	57
Globoquadrina altispira (CUSHMAN & JARVIS)																																			
Globoquadrina dutertrei (D'ORBIGNY)	32		14		14	16	13		21	19		6	12	10		X	14	9	26	4	8	15	25	X	17	6	3		26	17	25	16		27	8
Globorotalia acostaensis																																			
Globorotalia menardii (D'ORBIGNY)	38				+	+			14	1		+	+		X		1	1		2		2	28		3				16					46	
Globorotalia punctulata D'ORBIGNY			14		1	5	2	3	+	2	3	+	2	2			9	3	15	5	14	3			5	1	1	X	5	6	2			1	
Globorotalia scitula (BRADY)					1			9				6	1						1			2			2	+	1								
Globorotalia truncatulinoides (D'ORBIGNY)			1		1	1	1		1				1				1		2		2				1	6		X		2					1
Orbulina bilobata (D'ORBIGNY)																						+							+						
Orbulina universa D'ORBIGNY	1		1		+	+				+		2	+					+				+	+		8	+			+	1	2	+		X	
Pulleniatina obliquiloculata (PARKER & JONES)	2				1				3	+										1		+	7		2				1	2		5		1	
Pulleniatina primalis BANNER & BLOW																																			
Sphaeroidinella dehiscens (PARKER & JONES)	4				+				+									+					1		+				+					2	
Sphaeroidinella subdehiscens BLOW																																			
BENTHONIC FORAMINIFERIDA (agglutinated) (%)	1	0	0		6	1	0	0	5	0	0	0	0	0	0	0	0	0	0	0	0	0	6	0	5	0	0	0	11	5	0	0	0	11	+
Adercotryma glomerata (BRADY)																																			
Cribrostomoides subglobosum (SARS)																													1					8	
Cyclammina trullisata (BRADY)	1																						2						1	1				1	
Cystammina galeata (BRADY)																																		+	
Eggerella bradyi (CUSHMAN)					4	+			2														1		2				2	4				1	+
Gaudryina atlantica (BAILEY)																																			
Glomospira charoides (JONES & PARKER)																																			
Hormosinidae																							1						+						
Hyperammina elongata BRADY																																			
Karreriella apicularis (CUSHMAN)																													1						
Martinottiella communis (D'ORBIGNY)																																			
Recurvoides turbinatus (BRADY)									1																										
Rhabdammina abyssorum SARS									1																										
Saccammina sphaerica BRADY									+																										
Sigmoilopsis schlumbergeri (SILVESTRI)													+																						
Siphotextularia rolshauseni PHLEGER & PARKER									1																3				3						
Siphotextularia spec. indet.						1																													
Textulariidae					2																		2												
Trochammina globigeriniformis PARKER & JONES																													3					+	
BENTHONIC FORAMINIFERIDA (calcareous) (%)	99	100	100		94	99	100	98	95	22	19	28	25	24	100	100	35	20	99	15	98	37	94	100	95	100	100	100	89	95	100	100	28	89	100
Amphicoryna scalaris (BATSCH)																																			
Anomalina globulosa CHAPMAN & PARR																																			
Anomalinoides io (CUSHMAN)									1														+			1			2		3	3			
Astrononion stelligerum (D'ORBIGNY)	3				2				1				+	+						1															
Bolivina albatrossi CUSHMAN																																			
Bolivina goësi CUSHMAN								2		2	2	1	1	2			2	2		2		4													
Bolivina pseudoplicata HERON-ALLEN & EARLAND																																			
Bolivinita quadrilatera (SCHWAGER)																																			
Brizalina alata (SEGUENZA)									+																										
Brizalina pettersoni (PARKER)								7	7	7	2	8	19	6			5	6		4		13													
Brizalina subaenariensis (CUSHMAN)								2	+	1	1		1	2			3	2		1		2													
Bulimina aculeata D'ORBIGNY										+	1		1				1	1																	
Bulimina alazanensis CUSHMAN																																			
Bulimina exilis BRADY												1	1				+	2		1		2													
Bulimina inflata SEGUENZA										+			+				1																		
Bulimina marginata D'ORBIGNY										1	1	1																							
Cassidella bradyi (CUSHMAN)					1	9	4					1							3		1				1	3	3			3	2			+	
Cassidella mexicana (CUSHMAN)									1																										
Cassidella spec. indet.																																			
Cassidulina carinata SILVESTRI								22	+	4	6	3		3			2	3		2		4													
Cassidulinoides bradyi (NORMAN)																																			
Cassidulinoides mexicana (CUSHMAN)																																			
Ceratobulimina pacifica CUSHMAN & HARRIS																																			
Chilostomella oolina SCHWAGER										+																									
Cibicides ex gr. lobatulus (WALKER & JACOB)																			1																
Cibicidoides robertsonianus (BRADY)	4				1				3										26				2		7				5			1		2	
Ehrenbergina trigona GOES																																			
Ehrenbergina undulata PARKER																																			
Epistominella exigua (BRADY)	27	29	28		18	3	4		15			1		3	16	16	1		1		4	1	9	X	7		19	X	9	2		4		8	9
Epistominella smithi (STEWART & STEWART)								29	1	2	3	1					1			1															
Eponides polius PHLEGER & PARKER																																			
Eponides pusillus PARR		49	30												72	35								X		7		X							19
Eponides regularis PHLEGER & PARKER									3																										
Euuvigerina peregrina (CUSHMAN)	+					+		2		2		3	+				2	1		1															
Francesita advena (CUSHMAN)																																			
Fursenkoina complanata (EGGER)								2																											
Gavelinopsis translucens (PHLEGER & PARKER)																																			
Globobulimina pacifica CUSHMAN												+	+									2													
Globocassidulina oblonga (REUSS)														+			1																		2
Globocassidulina subglobosa (BRADY)					5				1				+										3		2				5	1	1	1		1	
Gyroidina lamarckiana (D'ORBIGNY)					6				1			1											2		1				4			1		1	3
Gyroidina soldanii D'ORBIGNY	7				1	+			2								1						2		8				5	1		4		2	+
Heterolepa pseudoungeriana (CUSHMAN)										+											2														
Hoeglundina elegans (D'ORBIGNY)										+																									
Hyalinea balthica (SCHROETER)										1	1																								
Lagena s.l.	3	1			6	12		6	2			1		2	+		2	1	+	+	+	2	6				7		4	3	2	3		5	5
Laticarinina pauperata (PARKER & JONES)																																			
Lenticulina peregrina (SCHWAGER)																																			
Lenticulina s.l.	+					+			1	+	1						1			+	+		+		3	1	+				1			1	
Lingulina seminuda HANTKEN																																			
Melonis affinis (REUSS)	2				4				3	+		1									1		5		8		2		7	1		16		4	
Melonis pompilioides (WILLIAMSON)					2	5			1										26		12		4		2		19			5	8	6		3	2
Miliolidae		1																							+	2	1				1	4			
Nodosaria longiscata D'ORBIGNY																																			
Nodosaria (incl. Dentalina)					+		4		+																+						1				+
Nummoloculina irregularis (D'ORBIGNY)						4																					1				8				+
Nuttallides umboniferus (CUSHMAN)	25				12	3	11		26										14		2		26		13				13	9	1	25		30	1
Ophthalmidium pusillum (EARLAND)								2																											
Oridorsalis umbonatus (REUSS)	10	3	7		8	13			2	+			+	3		2	5		7		38		10		1	46	20		6	6	17	1	4	11	32
Osangularia culter (PARKER & JONES)														+																					
Planulina ariminensis D'ORBIGNY																																			
Planulina wuellerstorfi (SCHWAGER)	8		10		9	19	33	5	7					3		10	6	+	14		32		15		22	7	18		10	36	18	19		14	8
Pleurostomella alternans SCHWAGER																			1																
Polymorphinidae																																			
Praeglobobulimina pupoides (D'ORBIGNY)																																			
Pseudonodosaria torrida (CUSHMAN)																																			
Pullenia bulloides (D'ORBIGNY)	8				8	2	4		5												1		3		8		7		7	8	6	7	12	3	1
Pullenia quinqueloba (REUSS)			1		1	4			1							2	1				1		1		2	3	+				1	2	4		
Pyrgo lucernula (SCHWAGER)																																			
Pyrgo murrhyna (SCHWAGER)	+				4	18			+				+									+	5		9	1	3		6	5	16			3	+
Pyrgo serrata (BAILEY)																																			
Pyrgo spp. indet.									2	+		+		+																					
Pyrgoella sphaera (D'ORBIGNY)																																			
Pyrulina cylindroides (ROEMER)																																			
Quinqueloculina spp. indet.	+		4		3	7	25		3										7		1		1		1	16			5	15	14	2			1
Rotamorphina laevigata (PHLEGER & PARKER)																																			
Robertina bradyi CUSHMAN & PARKER																															+				
Saracenaria italica DEFRANCE																																			
Siphonodosaria spp. indet.												+																							
Siphouvigerina auberiana (D'ORBIGNY)								2		+			+																						
Sphaeroidina bulloides D'ORBIGNY																				+															
Spirillina decorata BRADY																																			
Spiroloculina spp. indet.										1		1	+	+				+		+		+													
Spirosigmoilina tenuis (CZJZEK)																																			
Trifarina angulosa (WILLIAMSON)												1																							
Triloculina tricarinata D'ORBIGNY																																			
Vaginulina spinigera BRADY																																			
Valvulineria arctica GREEN																																			
Valvulineria complanata (D'ORBIGNY)																						2													
SHALLOW WATER BENTHONIC FORAMINIFERIDA (%)	0	0	0		0	0	0	2	0	78	81	72	75	76	0	0	65	80	1	85	2	63	0	0	0	0	0	0	0	0	0	0	72	0	0
Ammonia beccarii (LINNE)								1		7	14	5	7	5			10	11		4		7											22		
Amphistegina lessonii D'ORBIGNY										2	4	5	4	2			1	3		4	1														
Asterigerinata mamilla (WILLIAMSON)																																			
Canceris oblongus (WILLIAMSON)													1	2			2			1		2													
Cibicides ex gr. lobatulus (WALKER & JACOB)										13	17	7	9	17			10	17		15		6											22		
Cribroelphidium gunteri (COLE)										1	1	2	2	3			5			3		4													
Cribrononion advenum (CUSHMAN)									+													1											4		
Cribrononion incertum (WILLIAMSON)																		1				2													
Dyocibicides spec. indet.										1			1	2			1	4		1															
Elphidiidae																																			
Elphidium crispum (LINNE)										+		1						+																	
Eponides repandus (FICHTEL & MOLL)										+	2	+	2	3				1		2		1													
Florilus asterizans (FICHTEL & MOLL)										8	4	9	5	3			15	6		4	+	6													
Florilus atlanticus (CUSHMAN)										2	3	3	1	1			2	2	1	2	+	2											4		
Hanzawaia ex gr. concentrica (CUSHMAN)																	1																8		
Miliolidae										2			1	1																					
Neoconorbina orbicularis (TERQUEM)									+	1	1																								
Nonionella spec. indet.				X				1		21	21	20	23	19			7	16		24	1	10													
Pararotalia armata (D'ORBIGNY)											1	2	3																						
Planorbulina mediterranensis D'ORBIGNY																																			
Poritextularia panamensis (CUSHMAN)										+		1	+					1		+		2													
Ptychomiliola separans (BRADY)										+		+						+				+													
Quinqueloculina spp. indet.										15	13	12	14	16			9	11		17		16													
Reussella miocenica CUSHMAN										+				+			2																		
Rosalina globularis D'ORBIGNY										4	1	5	1	1				2				3											4		
Textulariidae										1				1				5		2		1													
Number of Radiolaria per gramme sediment	3	7	13	3	8300	120	5	5500	17						860	+			210		13		520		540	1100	180		430	600	280	300	1	70	410
Echinoidea (spines)	6		+		3	+		15	3	480	19	650	550	+		+	1	150		730	1	770	3		1		1		14		7	1		3	1
Echinoidea (plates)										+		+					12	570				3800													
Mollusca				+				10	+	100	2	600	310	2			1	190		350	1	750				5					14		+	+	1
Ostracoda					+	+				10		4	14	+			+	30	+	16	1	+					+		1		1				+
Pisces (teeth)																			+																
Pisces (otoliths)						+	+											+	+		+										+				
Bryozoa										15	+		42	1						17		17													

Table 3

	NIGER SLOPE (BASIN)											NIGEF. SLOPE (UPHEAVAL)																	SEA MOUNT			
NELSON core	26		25			24			23			22			12				21				20			18			35			
sample interval (in cm from top)	4-14	95-103	5-15	74-82	170-178	0-8	74-82	171-179	4-12	100-108	170-178	10-16	117-123	200-206	17-20	100-103	143-146	230-233	3-9	59-65	156-162	250-256	40-46	99-105	205-211	45-51	138-144	236-242	20-28	90-98	150-158	246-254
water depth at sample station (metres)	337		689			692			1340			1630			1810				2430				2587			2799			2829			
dry weight of sample (grammes)	76	82	51	78	74	74	76	72	56	68	56	50	54	58	30	32	38	29	50	64	62	64	59	56	67	64	62	59	87	84	96	76
weight-percentage of residue	2	2	1	1	+	1	1	3	1	+	+	2	+	+	4	11	12	3	17	34	3	4	20	10	11	13	1	4	87	51	88	50
PLANKTONIC FORAMINIFERIDA																																
Candeina nitida D'ORBIGNY																										+						
Globigerina bulloides D'ORBIGNY			2		2		1	2	2			1			3						1	1	2	2	4	2	2	2	7	6		4
Globigerina inflata D'ORBIGNY			2			1			3								3			5				2						2	2	2
Globigerina pachyderma (EHRENBERG)							2														1	1	3	3	1					4		
Globigerinella aequilateralis (BRADY)																	3				1			1		2				+		
Globigerinita glutinata (EGGER)				1	1		1		1								1								2						3	
Globigerinoides conglobatus (BRADY)	+	+					+	+	+	+						+	+			+		+	+	1		+			+			
Globigerinoides extremus BOLLI & BERMUDEZ																																
Globigerinoides immaturus LEROY																																
Globigerinoides ruber (D'ORBIGNY) (pink)	28	37	39	41	24	34	35	33	30	35	40	22	35	44	26	29	23	41	+	23	21	33	4	8	11	12	17	12	25	5	2	6
Globigerinoides sacculiferus (BRADY)				3		4	1	+	+		1	1			7	+	1	3	3	+		1	3	+		4	1		10	4	2	2
Globigerinoides trilobus (REUSS)	18	27	29	21	39	21	21	29	30	29	30	45	31	28	25	33	27	24	26	36	30		38	35	28	33	34	28	12	28	20	30
Globoquadrina altispira (CUSHMAN & JARVIS)																																
Globoquadrina dutertrei (D'ORBIGNY)	28	26	8	14	18	24	16	8	15	22	18	9	15	17	12	19	19	13	26	14	10	8	11	29	19	22	18	22	9	20	17	12
Globorotalia acostaensis																																
Globorotalia menardii (D'ORBIGNY)	20	2	8	12	+	7	+	+	12	8	1	9	3	3	16				19	+	16	7			17	2		6	1	1	24	6
Globorotalia punctulata D'ORBIGNY			2	3	11	4	3	6	4	1	4	7	10	3	2	1	4	3	6	4	5	11	2	2	9	4		3	5	2	4	6
Globorotalia scitula (BRADY)									1									1		1	1	3		3	1				1			1
Globorotalia truncatulinoides (D'ORBIGNY)		1	2								3		1															1	1	1		1
Orbulina bilobata (D'ORBIGNY)																																
Orbulina universa D'ORBIGNY	+	+	+	1		+	3	1	+	2	1	1	+	1	1	1	2	+	2	5	3	4	3	+	1	5	4	2	1	1	2	+
Pulleniatina obliquiloculata (PARKER & JONES)	+	+	+	1		+		+	1			+		2	1	+			+	1	+	2	1		+	1		+	+		1	1
Pulleniatina primalis BANNER & BLOW																																
Sphaeroidinella dehiscens (PARKER & JONES)	+	1	+			+			1		+	+			+				+		+	+					1	+				+
Sphaeroidinella subdehiscens BLOW																									1							
BENTHONIC FORAMINIFERIDA (agglutinated) (%)	0	0	+	5	1	1	1	3	4	6	3	4	2	2	7	7	3	3	7	1	9	1	4	1	6	4	8	0	3	5	2	2
Adercotryma glomerata (BRADY)																																
Cribrostomoides subglobosum (SARS)																																
Cyclammina trullisata (BRADY)																			1													
Cystammina galeata (BRADY)																																
→ Eggerella bradyi (CUSHMAN)				2				1	2	3	1	1			5	3	1	3	4	1	7	+	3	1	3	3	3		2	4	2	2
Gaudryina atlantica (BAILEY)			+	+		+	+	+	1		1	2				+	+															
Glomospira charoides (JONES & PARKER)																																
Hormosinidae																													+			
Hyperammina elongata BRADY																																
Karreriella apicularis (CUSHMAN)																																
Martinottiella communis (D'ORBIGNY)				1																												
Recurvoides turbinatus (BRADY)																																
Rhabdammina abyssorum SARS																																
Saccammina sphaerica BRADY																																
Sigmoilopsis schlumbergeri (SILVESTRI)				2	1	+	1	2	1	3	1	1	2	2	2	3	2		2		2	1	1		3	1	5					
Siphotextularia rolshauseni PHLEGER & PARKER																																
Siphotextularia spec. indet.																				+										1		
Textulariidae																1													1	+	+	
Trochammina globigeriniformis PARKER & JONES																																
BENTHONIC FORAMINIFERIDA (calcareous) (%)	95	95	100	95	99	99	99	93	96	94	97	96	98	98	93	93	97	97	93	99	91	99	96	99	94	96	92	100	97	95	98	98
Amphicoryna scalaris (BATSCH)	2	2						1																								
Anomalina globulosa CHAPMAN & PARR																5	5	2	2	1	+	1	2	1		+		1	1			2
Anomalinoides io (CUSHMAN)					3		1	3				1			1								1				4	3		3		
Astrononion stelligerum (D'ORBIGNY)																				4			3	1		7				8		1
Bolivina albatrossi CUSHMAN																																
Bolivina goësi CUSHMAN																																
Bolivina pseudoplicata HERON-ALLEN & EARLAND																																
Bolivinita quadrilatera (SCHWAGER)																																
Brizalina alata (SEGUENZA)														2																		
Brizalina pettersoni (PARKER)																																
Brizalina subaenariensis (CUSHMAN)	1																															
Bulimina aculeata D'ORBIGNY	7	1					1	1			4		1	2		4	13	42		+	17	18		1	2	1	2	6				
Bulimina alazanensis CUSHMAN															+	1	2	2	2	6	7	2	2	4	4	4	1	3	7	7	10	4
Bulimina exilis BRADY																																
Bulimina inflata SEGUENZA			1			1		1	5	6	5	6	9	2	6	3	1	1		+		1		+	+	1	2	1				
Bulimina marginata D'ORBIGNY	3	26	11	12	10	7	8	3																								
Cassidella bradyi (CUSHMAN)																								1					+	2		
Cassidella mexicana (CUSHMAN)						+																										
Cassidella spec. indet.																																
Cassidulina carinata SILVESTRI	1	1	1	2	2	3	3	3	1		1		1	5	4	1	4	2	1	4	+		5	5		1	1			3		+
Cassidulinoides bradyi (NORMAN)																						1										
Cassidulinoides mexicana (CUSHMAN)																			+													
Ceratobulimina pacifica CUSHMAN & HARRIS				2	+	+	1	1	11	3	4	3	2		8	5	2	1	1		4											
Chilostomella oolina SCHWAGER							1																							1		
Cibicides ex gr. lobatulus (WALKER & JACOB)																									+				5	4	6	1
Cibicidoides robertsonianus (BRADY)			3	2		8	1	2	5		7	1			1	3	3	1	5	2		2	2	2		4					3	2
Ehrenbergina trigona GOËS																																4
Ehrenbergina undulata PARKER	17	22		1		1	3	7									1															
→ Epistominella exigua (BRADY)																				2					1				+	2	1	1
Epistominella smithi (STEWART & STEWART)																																
Eponides polius PHLEGER & PARKER			1																			+										
Eponides pusillus PARR															+																	
Eponides regularis PHLEGER & PARKER	1		1	2	2		2	2	1			4	1					1														
Euuvigerina peregrina (CUSHMAN)	20	25	46	28	27	47	26	25	19	10	20	21	32	6	8	2	2	6	23	20	22	46	23	33	35	24	13	33	3	7	4	5
Francesita advena (CUSHMAN)																																
Fursenkoina complanata (EGGER)																																
Gavelinopsis translucens (PHLEGER & PARKER)								1	1						3	1	+															
Globobulimina pacifica CUSHMAN		2	2		+	+	2	+					2	1							1						2	1				
Globocassidulina oblonga (REUSS)																																
→ Globocassidulina subglobosa (BRADY)		2													1	2	4	+	6	4	+		3		2	+		1	12	3	13	24
Gyroidina lamarckiana (D'ORBIGNY)																			1													
Gyroidina soldanii D'ORBIGNY			1					1	3	3	3		1	2	6	6	1	1	9	1	1	3	2	+	1	+	3	1	2	4	+	3
Heterolepa pseudoungeriana (CUSHMAN)	3	3	21	36	35	18	28	21	6	26	17	10	16	48	1	3	+		3		2	2			2	3		2				
Hoeglundina elegans (D'ORBIGNY)	+	1	2	2	5	2	3	+	2	+	+	1	1		1	6	5	+	11	10	17	2	27	3	12	10	4	6	3	10	3	
Hyalinea balthica (SCHROETER)							2	1																								
Lagena s.l.		2	1					1		2		1		1	2	1	2		2	2	1	2	3	2	3	3	1		2	5	4	5
Laticarinina pauperata (PARKER & JONES)									1							1		1										1	2		+	1
Lenticulina peregrina (SCHWAGER)						+						1	+																			
Lenticulina s.l.	6	1	+	1	1	+	1	2	+	1		2			1	+		1				1	1		+	+	2	+	1		1	+
Lingulina seminuda HANTKEN				+	+	+																										
Melonis affinis (REUSS)	1		1			2	2		2			2	2		+	1	+	2	1	8	1	3	4	17	1	6	4	9		+		+
→ Melonis pompilioides (WILLIAMSON)																				+	+		2	5		5	6	6		1	+	
Miliolidae										1			+	2	1	1	+	1														
Nodosaria longiscata D'ORBIGNY																																
Nodosaria (incl. Dentalina)		1				1	+	1				+						1	+				1	+		+				1	+	1
Nummoloculina irregularis (D'ORBIGNY)		+						+				1		2	2		1	+							+							
→ Nuttallides umboniferus (CUSHMAN)																													12	11	1	12
Ophthalmidium pusillum (EARLAND)																																
→ Oridorsalis umbonatus (REUSS)	1		2		+	2	5	3	6	5	6	2	5	2	10	3	1	2	4	6	4	5	2	7	13	4	9	8	9	4	7	8
Osangularia culter (PARKER & JONES)									4	6	4	4		1	4	8	11	4	1	2	+	+	1		1				2			+
Planulina ariminensis D'ORBIGNY	3	2			+		+	1																								
→ Planulina wuellerstorfi (SCHWAGER)				2				1	4	7	4	3	8	7	2	1	3	2	3	3	6	3	2	2	7	+	15	2	1	1	3	+
Pleurostomella alternans SCHWAGER																																+
Polymorphinidae																		1														
Praeglobobulimina pupoides (D'ORBIGNY)					2		2			1		1				1						+		+		2						
Pseudonodosaria torrida (CUSHMAN)				1	+		+																						1			+
→ Pullenia bulloides (D'ORBIGNY)			1					1	5	2	2	7	1	1	4	5	2	4	1	3	1	1	2	5	4	3	5	7	1	3	5	+
Pullenia quinqueloba (REUSS)				1									1		1	1			3		+	+	1	2	2	4	1		7	1	1	1
Pyrgo lucernula (SCHWAGER)									+						+	+		+				1				+						
Pyrgo murrhyna (SCHWAGER)									+	+				1	4	2	1	2	1	1	+	+	3	+	1	2	1	1	6	12	7	5
Pyrgo serrata (BAILEY)																														+		
Pyrgo spp. indet.		+		+			+	+					+				+	+		+												
Pyrgoella sphaera (D'ORBIGNY)											2	1				1																
Pyrulina cylindroides (ROEMER)																																
Quinqueloculina spp. indet.		1	+	+	1	+	1	+		5	2	+			4	2	+	+				1	1	+		1	1					1
Rotamorphina laevigata (PHLEGER & PARKER)																									1							
Robertina bradyi CUSHMAN & PARKER				1											+																	
Saracenaria italica DEFRANCE	+	+		+				+																								
Siphonodosaria spp. indet.																																1
Siphouvigerina auberiana (D'ORBIGNY)									8	4	5	1			14	22	31	10	13	15		3	3	1	+	3	8	2	18	4	17	8
Sphaeroidina bulloides D'ORBIGNY			3	2	9	4	5	1	11	11	6	23	15	12	+	+	+	1	+	2	2	1	+	4	+	5	6	2	+	1	3	+
Spirillina decorata BRADY																																
Spiroloculina spp. indet.					+	+	1	4			+			+				1														
Spirosigmoilina tenuis (CZJZEK)																																
Trifarina angulosa (WILLIAMSON)	25	3																														
Triloculina tricarinata D'ORBIGNY																															1	
Vaginulina spinigera BRADY									+					1																		
Valvulineria arctica GREEN																				+			+									
Valvulineria complanata (D'ORBIGNY)																				+												
SHALLOW WATER BENTHONIC FORAMINIFERIDA (%)	5	5	0	0	0	0	0	4	0	0	0	0	0	0	0	0	0	0	0	0	0	0	0	0	0	0	0	0	0	0	0	0
Ammonia beccarii (LINNE)																																
Amphistegina lessonii D'ORBIGNY																																
Asterigerinata mamilla (WILLIAMSON)																																
Cancris oblongus (WILLIAMSON)	2	2						1																								
Cibicides ex gr. lobatulus (WALKER & JACOB)																																
Cribroelphidium gunteri (COLE)																																
Cribrononion advenum (CUSHMAN)																																
Cribrononion incertum (WILLIAMSON)																																
Dyocibicides spec. indet.																																
Elphidiidae																																
Elphidium crispum (LINNE)																																
Eponides repandus (FICHTEL & MOLL)																																
Florilus asterizans (FICHTEL & MOLL)																																
Florilus atlanticus (CUSHMAN)																																
Hanzawaia ex gr. concentrica (CUSHMAN)	3	3						3																								
Miliolidae																																
Neoconorbina orbicularis (TERQUEM)																																
Nonionella spec. indet.																																
Pararotalia armata (D'ORBIGNY)																																
Planorbulina mediterranensis D'ORBIGNY																																
Poritextularia panamensis (CUSHMAN)																																
Ptychomiliola separans (BRADY)																																
Quinqueloculina spp. indet.																																
Reussella miocenica CUSHMAN																																
Rosalina globularis D'ORBIGNY																																
Textulariidae																																
Number of Radiolaria per gramme sediment			7	9					80	120	1	740	90	10	1300	3	67	73	520	150	16	15	140	220	4	73	+		2300	2300	690	
Echinoidea (spines)	34		1		3		10			3	2		+	1	5	10	13	23	6	5	+	10	17	70	16	18	1	12	72	14	6	7
Echinoidea (plates)				+			1	+	+		+		+	7																		
Mollusca	+	210		2	4	6	43	1		+	27	+			30	450	490	67		2									+			
Ostracoda	+			+	+	+	2	2			+	+	+	+	22	1	45	+	+	20	1	+	5	+	5	10			12	+		1
Pisces (teeth)	7											+			+											1						
Pisces (otoliths)		3	+	+	+		+	1	+		+	+			+	2	1	+		+		+	+	10	+	+						
Bryozoa																																

Table 4

	CALABAR CANYON							PRINCIPE CHANNEL																												
NELSON core	15				14			27			29				28		32			34			37			36		39				40		41		
sample interval (in cm from top)	25-29	108-111	164-167	253-256	16-19	102-105	172-175	7-15	62-70	100-104	5-13	52-60	103-107	212-220	35-43	67-71	0-8	36-40	100-106	16-24	110-118	200-208	3-11	46-54	74-78	10-18	35-39	3-13	31-38	72-80	132-142	20-28	110-116	23-29	55-61	140-146
water depth at sample station (metres)	1441				1497			2552			3129				3240		3509			3846			4430			4540		4617				4635		4635		
dry weight of sample (grammes)	29	28	30	36	22	26	24	42	74	90	65	71	90	79	70	90	57	87	82	69	55	56	82	47	96	70	103	42	112	30	50	52	37	61	160	160
weight-percentage of residue	1	2	1	3	3	4	3	3	33	33	15	13	32	33	35	31	28	43	24	22	15	13	11	4	68	33	41	15	33	2	3	11	7	2	38	38
PLANKTONIC FORAMINIFERIDA																																				
Candeina nitida D'ORBIGNY																																				
Globigerina bulloides D'ORBIGNY	3	3	2	1	1	2	4	3	1		1	3		2	4	2	6	1	2		5	4	5	5	7	2		3	2	5	4	3	5			
Globigerina inflata D'ORBIGNY													1	2			4				3	3			1		3	10				2	3		1	3
Globigerina pachyderma (EHRENBERG)												2	6								1	1			1		1					11	9	13	5	9
Globigerinella aequilateralis (BRADY)		1			1									1			1	+		1						3			+							
Globigerinita glutinata (EGGER)								1			5		1	2			2			1	4	2		1	1	1		2	2			+	8			
Globigerinoides conglobatus (BRADY)											+	+	+	1	+	+				1	+	1	+				+	+	+			+	+		+	
Globigerinoides extremus BOLLI & BERMUDEZ																																				
Globigerinoides immaturus LEROY																																				
Globigerinoides ruber (D'ORBIGNY) (pink)	50	52	47	52	59	66	76	41	51	61	49	25	12	38	45	21	20	34	39	15	24	28	4	16	57	6	17		20	8	32	3	7	4	17	29
Globigerinoides sacculiferus (BRADY)		1	1	1		1	1	1	1		4	5	1	1	2	8	1	3		19	2	3	4	5		3	2		2			3	+	+	2	1
Globigerinoides trilobus (REUSS)	10	16	16	17	14	2	4	20	10	12	9	45	22	6	16	18	17	30	18	10	7	16	18	38	34	16	24	14	32	47	28	25	12	28	25	35
Globoquadrina altispira (CUSHMAN & JARVIS)																																				
Globoquadrina dutertrei (D'ORBIGNY)	12	9	10	10	7	13	5	17	7	13	14	14	18	27	15	31	24	12	20	30	22	22	38	23	1	25	17	37	18	14	18	22	21	24	16	18
Globorotalia acostaensis																																				
Globorotalia menardii (D'ORBIGNY)	2	+			5	5		6	5	+	8		2	+	3	+	10	1	1	1			11		2	10	4	23	2	17	+	+		13	3	2
Globorotalia punctulata D'ORBIGNY	8	2	3	2	7	9	6	5	4	1	1	2	7	5	7	5	5	3	5	6	3	4	5	2	4	2	3	2	2	7	7	4	3	2	3	3
Globorotalia scitula (BRADY)			1					1			1	1	1	1		1									1					1		1				1
Globorotalia truncatulinoides (D'ORBIGNY)															1									2		1	2	1				2		1	+	
Orbulina bilobata (D'ORBIGNY)																																				
Orbulina universa D'ORBIGNY	2	1	1	3	2	2	+	1		2		1	+	1	+	1	2	3	2	+	+	1	+	1	1	2	1	+	2	+		+	+		+	+
Pulleniatina obliquiloculata (PARKER & JONES)					+			1	2		1	+	+	+	+	+	1			1		+	1			1	+	2	+			3		+	+	+
Pulleniatina primalis BANNER & BLOW																																				
Sphaeroidinella dehiscens (PARKER & JONES)	+							+	+		+				+	+	+	1	+	+			+			+	1	+		1		+		1	1	+
Sphaeroidinella subdehiscens BLOW																																				
BENTHONIC FORAMINIFERIDA (agglutinated) (%)	2	4	4	2	4	5	1	4	2	1	2	6	1	2	2	0	8	1	0	4	7	4	8	3	1	7	1	4	0	2	3	3	0	1	0	0
Adercotryma glomerata (BRADY)																																				
Cribrostomoides subglobosum (SARS)																							2													
Cyclammina trullisata (BRADY)																										2		2								
Cystammina galeata (BRADY)																																				
Eggerella bradyi (CUSHMAN)	1	2	2		3	1		1	1			3		1	1	+	1	1	+	2	7	3	4	3	1	2		1		2		2		1		
Gaudryina atlantica (BAILEY)		1	1	+	+		+																													
Glomospira charoides (JONES & PARKER)																	1																			
Hormosinidae																																				
Hyperammina elongata BRADY																	+																			
Karreriella apicularis (CUSHMAN)																																				
Martinottiella communis (D'ORBIGNY)																																				
Recurvoides turbinatus (BRADY)																										2						1				
Rhabdammina abyssorum SARS																	2						1													
Saccammina sphaerica BRADY						+			+			+				+																				
Sigmoilopsis schlumbergeri (SILVESTRI)	1	1	1	1	1	4	1	1	1	1	2	2	1			+	1										1				3					+
Siphotextularia rolshauseni PHLEGER & PARKER														1			1					1				1										
Siphotextularia spec. indet.															1					1								1								
Textulariidae																	1			1			1													
Trochammina globigeriniformis PARKER & JONES																																				
BENTHONIC FORAMINIFERIDA (calcareous) (%)	98	96	96	92	96	95	99	96	90	33	98	94	31	86	96	64	92	72	98	96	93	96	92	97	80	91	58	96	55	94	49	97	99	99	39	52
Amphicoryna scalaris (BATSCH)																																				
Anomalina globulosa CHAPMAN & PARR		1						1				+				1		+		1						1	1									
Anomalinoides io (CUSHMAN)																+	3						2													
Astrononion stelligerum (D'ORBIGNY)								1	1		4	6	1	5	3		4		2	6	4	7			1	2		3	1	2	1	3		3		6
Bolivina albatrossi CUSHMAN																																				2
Bolivina goësi CUSHMAN										3			4	1		4		5	5						3		6		5				2		10	8
Bolivina pseudoplicata HERON-ALLEN & EARLAND																																				
Bolivinita quadrilatera (SCHWAGER)	1	1	1				1		1									2	1						2		1						1	1		1
Brizalina alata (SEGUENZA)	1		5	3										2	1			5	1						12		4		3		4		5		2	4
Brizalina pettersoni (PARKER)																																				
Brizalina subaenariensis (CUSHMAN)									1	1				2		2		4							3		1		3		2		1		2	2
Bulimina aculeata D'ORBIGNY				14										2		2		4	16						5		6		4		2				3	1
Bulimina alazanensis CUSHMAN			1	1					1		1	1	1		1																				+	
Bulimina exilis BRADY													1		1																					
Bulimina inflata SEGUENZA	4	5	6	6	12	5	8	1					1			1		+	1								1									
Bulimina marginata D'ORBIGNY								1		1			1			1									1				2		1					+
Cassidella bradyi (CUSHMAN)																	+				4	1				2										
Cassidella mexicana (CUSHMAN)														1					1						1											+
Cassidella spec. indet.																																				
Cassidulina carinata SILVESTRI	5				5	3	1	2	3	1		1	5	3	2	6		6	7					1	13		6		4		3		2		7	5
Cassidulinoides bradyi (NORMAN)																			1																	
Cassidulinoides mexicana (CUSHMAN)				13									1	1				1	2						2		1		2		2				+	+
Ceratobulimina pacifica CUSHMAN & HARRIS	12	9	3		10	13	13	2		1						1																				
Chilostomella oolina SCHWAGER		1								+								1	1	5	1							1							1	1
Cibicides ex gr. lobatulus (WALKER & JACOB)																																				
Cibicidoides robertsonianus (BRADY)	1	13	1	4		1	3		6		16	1			3	1	9	1	2	3	1						1	3				7		3	+	
Ehrenbergina trigona GOËS																																				
Ehrenbergina undulata PARKER																																				
Epistominella exigua (BRADY)											2	1					1	1	1	2	7	5	9	11	+	11		13		7		18	10	27	+	
Epistominella smithi (STEWART & STEWART)				+					2	4		1	4	2	2	3		5			1				12		8		3		4		7		4	6
Eponides polius PHLEGER & PARKER																																				
Eponides pusillus PARR																																				
Eponides regularis PHLEGER & PARKER																1		1																		
Euuvigerina peregrina (CUSHMAN)	11	13	20	9	7	11	11	32	16	7	11	8	3	13	22	9	3	11	14				1	1	7		7		5		4		2		1	1
Francesita advena (CUSHMAN)																								1		1						1				
Fursenkoina complanata (EGGER)																																				
Gavelinopsis translucens (PHLEGER & PARKER)			1	+	4	1	+		+	1								+																		
Globobulimina pacifica CUSHMAN		1	1	2		1			4			+	2	2		+			2	1		1			3				2		1				+	1
Globocassidulina oblonga (REUSS)																																				
Globocassidulina subglobosa (BRADY)								2	3						+	1	1	+					1	1		5		4				1	2	1		+
Gyroidina lamarckiana (D'ORBIGNY)																							3	1				4	1			4		3		
Gyroidina soldanii D'ORBIGNY	2	2	3	2	1	2	6	1		1	3	2			3	+	2	1	1	12	1	1	3	1		3	1	1		5	2	3		2	+	
Heterolepa pseudoungeriana (CUSHMAN)	14	13	17	10	5	8	18	1	+	5		1	3	+	3	5	3	2	5				4	1			1		1		1				2	4
Hoeglundina elegans (D'ORBIGNY)	2	5		+	1	3	2	16	17	1	11	13	1		11	4	7	4	7	4	1						1		2						+	+
Hyalinea balthica (SCHROETER)										1			2			1		1	1						+						1					+
Lagena s.l.	2	2	3	1	1	2		3	1		2	2		2	2	1	7	2	2	5	5	4	5	11		12		10		7	1	2		2		
Laticarinina pauperata (PARKER & JONES)	2	2	1		2	2	1									1	1		+														4			
Lenticulina peregrina (SCHWAGER)		1																+																		
Lenticulina s.l.	+			1	+	+	1	1	1	+	2	1	+		1	1	2	1		1	1	1	1	1			2	1	3		2	3	1	2		1
Lingulina seminuda HANTKEN																																				
Melonis affinis (REUSS)	1				1		1	1	3		1	1	+	10		1		1	1		9	2	5		1	2	1	3	1		2	7	2	2		+
Melonis pompilioides (WILLIAMSON)											2	6	1	3	4	1	3	1	2	1	13	5		6	2		1	1	+	2	1	4	9		+	1
Miliolidae			3	1		3	1	1	4			+	+	1	1	1		+				1														
Nodosaria longiscata D'ORBIGNY																																				
Nodosaria (incl. Dentalina)					1	1	1	1		+	+		+	1	+	+	+		1	+				2									+		+	
Nummoloculina irregularis (D'ORBIGNY)		1	2	1	2	1		3	5			1		2	1	2			1																	
Nuttallides umboniferus (CUSHMAN)																							21	4	2	16	1	14		28	1	8	10	10		+
Ophthalmidium pusillum (EARLAND)																																				
Oridorsalis umbonatus (REUSS)	9	4		4	4	4	1	4	2	1	16	7		5	10	1	10	2	3	18	21	12	11	16		9	1	9	4	18		15	14	27	2	2
Osangularia culter (PARKER & JONES)	6	6	5	4	3	4	9							1		1		1									1		1		1				+	1
Planulina ariminensis D'ORBIGNY																			+																	
Planulina wuellerstorfi (SCHWAGER)	6	4	5		5	9	6	10	10	1	10	11	+	10	8	4	21	1	4	23	9	15	12	11	3	11	1	10		18	4	7	14	6	3	1
Pleurostomella alternans SCHWAGER																																				
Polymorphinidae					1				1																						2					
Praeglobobulimina pupoides (D'ORBIGNY)	1					1	1				1	2		7	10	+		3	1		1	+							1							1
Pseudonodosaria torrida (CUSHMAN)																		+																		
Pullenia bulloides (D'ORBIGNY)	2	3		1	4	3	3	5	2	2	4	2		1	3	1	4	1	3	1	1	2	3	9	3	6	1	10		5	1	8	6	6	+	
Pullenia quinqueloba (REUSS)		1		1				1	1		8	4		1			3			4			1	2		1	1	1				2				+
Pyrgo lucernula (SCHWAGER)														+											1											
Pyrgo murrhyna (SCHWAGER)	1		1	2	4	2		3	2		2	4		4	1	2	6	+	1	7	11	3	5	6	1	2		2	2	2		3	2	1	+	1
Pyrgo serrata (BAILEY)														+		+	+																			
Pyrgo spp. indet.					2							+		+	+	+		1	+		+			1					+		1				+	
Pyrgoella sphaera (D'ORBIGNY)																		+	2																	+
Pyrulina cylindroides (ROEMER)																								1					1							
Quinqueloculina spp. indet.	1	+	+	3	6	2		1			+	6		+	3		2		2	2			4	2		7		2				4	2	1		
Rotamorphina laevigata (PHLEGER & PARKER)																																				
Robertina bradyi CUSHMAN & PARKER	1	1			+	1			1					1				+																		
Saracenaria italica DEFRANCE																																				
Siphonodosaria spp. indet.																			1																	
Siphouvigerina auberiana (D'ORBIGNY)	2	1	1	1	1	2	4	2	1			5		2		1		2			1	32	1	7	1		1		1				1		+	1
Sphaeroidina bulloides D'ORBIGNY	11	5	9	6	6	10	6			1						1	+	1	+								1	1	+						1	
Spirillina decorata BRADY									1																											
Spiroloculina spp. indet.			+	1			1	+	+							+			1						1		+		+		1			1	+	+
Spirosigmoilina tenuis (CZJZEK)																																				
Trifarina angulosa (WILLIAMSON)																																				
Triloculina tricarinata D'ORBIGNY					2																1															
Vaginulina spinigera BRADY																																				
Valvulineria arctica GREEN														1		1																				
Valvulineria complanata (D'ORBIGNY)				+																																
SHALLOW WATER BENTHONIC FORAMINIFERIDA (%)	0	0	0	6	0	0	0	0	8	66	0	0	68	12	2	36	0	27	2	0	0	0	0	0	19	2	41	0	45	4	48	0	1	0	61	48
Ammonia beccarii (LINNÉ)				6						7			11	3	1	8		7	1						4	1	3		8		9		1		7	5
Amphistegina lessonii D'ORBIGNY										6			2		+			1	+										3		6				2	3
Asterigerinata mamilla (WILLIAMSON)																																			+	
Cancris oblongus (WILLIAMSON)													2			+		2							+											
Cibicides ex gr. lobatulus (WALKER & JACOB)													1			1		2							+		6				4				5	
Cribroelphidium gunteri (COLE)																5		1									2									
Cribrononion advenum (CUSHMAN)										2			+		1	1											1									
Cribrononion incertum (WILLIAMSON)									1																											
Dyocibicides spec. indet.																																				
Elphidiidae									1	2			2	1											3				4						6	
Elphidium crispum (LINNÉ)																																				
Eponides repandus (FICHTEL & MOLL)										1			2			1		1									1			2						1
Florilus asterizans (FICHTEL & MOLL)																																				
Florilus atlanticus (CUSHMAN)									1				1	1					1						+	1			2	2	2				2	2
Hanzawaia ex gr. concentrica (CUSHMAN)										1			3	1		2		1									2				5				1	2
Miliolidae										13																			1							
Neoconorbina orbicularis (TERQUEM)																											1								1	1
Nonionella spec. indet.																																				
Pararotalia armata (D'ORBIGNY)									4	26			22	1		7		6							3		18		12		14				18	19
Planorbulina mediterranensis D'ORBIGNY																																				
Poritextularia panamensis (CUSHMAN)																+																				
Ptychomiliola separans (BRADY)																+													2							+
Quinqueloculina spp. indet.													13	5		11		5							5		4		8		8				14	9
Reussella miocenica CUSHMAN																		1											2							2
Rosalina globularis D'ORBIGNY									1																											
Textulariidae										7			4												1		1		3						2	3
Number of Radiolaria per gramme sediment	470	47			620	580	900	3200	860	1	3800	3100	61	1200	3600	17	2400	190	150	1700	900	2000	19000	6100	150	14000	21	1500	2	49	+	9200	1700			
Echinoidea (spines)	+	1	7	5		1		+	4	4	8	4	23	3	9	80	4	15	12	15	+	7	1		46	3	16	1	9	+	2	.		2	8	8
Echinoidea (plates)			1								+					+		+	+						3		2		+				+			
Mollusca	21	380	290	70	200	60	270	140	110	31	6	13	55	3	16	230		110	16		+				140		37		35		1				41	38
Ostracoda	+	1	+	+	3			1	1	+	+	1	2	1	1	8	1	7	+		+	1	+	+	41	+	2		6	+	+			+	1	3
Pisces (teeth)		1						1	+	+	+	3	+	+					+		+	+					+								+	+
Pisces (otoliths)										+			2																+							7
Bryozoa																																				

Table 5

	GABON CONT. RISE				NIGER RISE																											
NELSON core	38				13			17				19			33			11				10				44			43			
sample interval (in cm from top)	6-14	97-105	170-178	240-248	25-30	107-111	178-182	11-15	110-114	187-191	256-259	10-18	80-87	150-158	8-16	40-48	150-158	5-8	71-74	127-130	231-234	8-11	35-38	87-91	173-177	28-32	110-115	180-186	28-34	100-106	186-192	244-250
water depth at sample station (metres)	3252				3200			3338				3626			3876			3897				4257				4479			4658			
dry weight of sample (grammes)	75	62	58	62	34	56	29	30	37	31	29	43	61	45	83	57	49	30	43	21	31	28	37	17	28	42	72	39	71	38	41	51
weight-percentage of residue	15	14	13	10	12	23	5	5	2	3	2	12	6	5	15	15	4	17	4	1	+	9	1	3	1	11	1	+	5	8	+	15
PLANKTONIC FORAMINIFERIDA																																
Candeina nitida D'ORBIGNY								1				+	+						1													
Globigerina bulloides D'ORBIGNY	3	2			6	7	5	2		1	1	4	6	4	2		4			1		3	1	7					6	5		
Globigerina inflata D'ORBIGNY	7		12	7	2		5			3	12			6	4	3	6			15	10	2		3	1			1	2	6	2	2
Globigerina pachyderma (EHRENBERG)		8	9	5			2			3	2		2	2	1		1											9	10	3	3	14
Globigerinella aequilateralis (BRADY)	1		+						1	1		1															1					
Globigerinita glutinata (EGGER)	2		1	1	+		5	2				+			1		7					1		+	2		1			5		6
Globigerinoides conglobatus (BRADY)		+	+		+		+	+		+	+	+	+	1	+	1	+		+	+				+			+	+	+	+	+	1
Globigerinoides extremus BOLLI & BERMUDEZ																																
Globigerinoides immaturus LEROY																																
Globigerinoides ruber (D'ORBIGNY) (pink)	13	13	14	9	24	13	30	18	28	30	18	17	11	4	13	15	24	9	19	10	10	3	20	17	13	2	4	17		17	3	6
Globigerinoides sacculiferus (BRADY)	2	4	+	4	2	4	1	4	+		2	3	6	+	5	7	1	6	3	5		3		2	1	2	4		2	2	+	2
Globigerinoides trilobus (REUSS)	26	15	10	15	25	45	14	20	36	11	14	26	42	34	32	43	4	22	33	17	33	7	34	13	32		60	40	16	14	10	14
Globoquadrina altispira (CUSHMAN & JARVIS)																																
Globoquadrina dutertrei (D'ORBIGNY)	26	24	34	37	12	13	25	17	13	19	14	18	22	18	19	19	20	19	18	11	17	32	15	26	14	44	15	6	44	17	53	5
Globorotalia acostaensis																																
Globorotalia menardii (D'ORBIGNY)	5				10			8	1			13	2		9	+		16	+			31	7			40	2		2			40
Globorotalia punctulata D'ORBIGNY	3	3	2	5	6	1	+	3	3	4	4	2	1	7	4		2		2	5	4		3	1	2		1		2	1	3	
Globorotalia scitula (BRADY)			1							2					1						1		1	1								3
Globorotalia truncatulinoides (D'ORBIGNY)			1	3					1										1										1	1		
Orbulina bilobata (D'ORBIGNY)										1																						
Orbulina universa D'ORBIGNY	1	+	1	2	2	1	+	2	+	2	2	2	2	+	1	1	1	+	1	1		1		1			+			1		1
Pulleniatina obliquiloculata (PARKER & JONES)	2	+			+	7		1	+			+	2		+	1		2	1			2	1			3			5		2	
Pulleniatina primalis BANNER & BLOW																																
Sphaeroidinella dehiscens (PARKER & JONES)	+				+			+				+			1			+	+			1	+			2	+		+			
Sphaeroidinella subdehiscens BLOW																																
BENTHONIC FORAMINIFERIDA (agglutinated) (%)	7	4	6	3	4	3	2	2	4	+	6	5	1	+	8	7	1	27	+	5	0	10	1	4	1	1	+	0	3	4	2	1
Adercotryma glomerata (BRADY)																		5														
Cribrostomoides subglobosum (SARS)															1																	
Cyclammina trullisata (BRADY)			1									1						5														
Cystammina galeata (BRADY)																																
Eggerella bradyi (CUSHMAN)	2	3	5	2	1	1	1		1	+	6	3	1	+	2	7	+	4	+	5		2	1	4	1	1	+		2	3	2	1
Gaudryina atlantica (BAILEY)								+																								
Glomospira charoides (JONES & PARKER)																		+														
Hormosinidae				+											1			3														
Hyperammina elongata BRADY																													+			
Karreriella apicularis (CUSHMAN)																																
Martinottiella communis (D'ORBIGNY)		+	1	1		1	1		1	+																						
Recurvoides turbinatus (BRADY)					1										1			+				5										
Rhabdammina abyssorum SARS																		4				3										
Saccammina sphaerica BRADY																																
Sigmoilopsis schlumbergeri (SILVESTRI)	5	1			2			2	2			1																				
Siphotextularia rolshauseni PHLEGER & PARKER						1												+														
Siphotextularia spec. indet.															1		1												1	1		
Textulariidae																																
Trochammina globigeriniformis PARKER & JONES															1			6														
BENTHONIC FORAMINIFERIDA (calcareous) (%)	93	96	94	97	96	97	98	98	96	100	94	95	99	100	92	93	99	73	100	95	100	90	99	96	99	99	100	100	97	96	98	99
Amphicoryna scalaris (BATSCH)																																
Anomalina globulosa CHAPMAN & PARR		2		1		+										2							+									
Anomalinoides io (CUSHMAN)	3				2	1	+	1	3	+	2	5	6	2	6		1	7	10	2		7	1	1		+	7	7	+	+		5
Astrononion stelligerum (D'ORBIGNY)	3	11	9	18	3	5	9	4	9	8	6	3	6	2	4	6	1	5	2								3		2			
Bolivina albatrossi CUSHMAN																																
Bolivina goësi CUSHMAN																										1						
Bolivina pseudoplicata HERON-ALLEN & EARLAND																																
Bolivinita quadrilatera (SCHWAGER)																																
Brizalina alata (SEGUENZA)				1																						1	1					
Brizalina pettersoni (PARKER)																										1	1					
Brizalina subaenariensis (CUSHMAN)						1													1								1					
Bulimina aculeata D'ORBIGNY					1	2																										
Bulimina alazanensis CUSHMAN				1	1	1																										
Bulimina exilis BRADY																																
Bulimina inflata SEGUENZA						1																				1						
Bulimina marginata D'ORBIGNY						+																										
Cassidella bradyi (CUSHMAN)		2	2	2	2	+	1										1			5				5						1	1	
Cassidella mexicana (CUSHMAN)																										1	+					
Cassidella spec. indet.																																
Cassidulina carinata SILVESTRI					1	4													1							1	1					
Cassidulinoides bradyi (NORMAN)																																
Cassidulinoides mexicana (CUSHMAN)																																
Ceratobulimina pacifica CUSHMAN & HARRIS						1																										
Chilostomella oolina SCHWAGER			1			+											+															
Cibicides ex gr. lobatulus (WALKER & JACOB)						+									1															1		
Cibicidoides robertsonianus (BRADY)	7	1			10	5		10	8	+		8	2	1	1			5	8				1	1		3	4		12			
Ehrenbergina trigona GOES																																
Ehrenbergina undulata PARKER																	+															
Epistominella exigua (BRADY)	3	1		1	1	3		1	3			2	2	2	1	4	1	+	1		23	4	12	6	23	23	22	18	11	18	29	23
Epistominella smithi (STEWART & STEWART)					1	6													2								1					
Eponides polius PHLEGER & PARKER																											+		+			
Eponides pusillus PARR																					29		19		37							+
Eponides regularis PHLEGER & PARKER																																
Euuvigerina peregrina (CUSHMAN)	7	5	1	10	14	7	20	23	10	15	17	4	2	16		1	38	+	2	2								1				+
Francesita advena (CUSHMAN)																																
Fursenkoina complanata (EGGER)																					2		4		1							
Gavelinopsis translucens (PHLEGER & PARKER)																																
Globobulimina pacifica CUSHMAN		1				3	7	2		4	6						1				1											
Globocassidulina oblonga (REUSS)																																
Globocassidulina subglobosa (BRADY)									1				2	2	1		3		1		1			1			1	1		1	2
Gyroidina lamarckiana (D'ORBIGNY)																					9	3	7		1	3	3		3			
Gyroidina soldanii D'ORBIGNY	2	3	4		1	1		2	+	+		3	5		5	7	+	6	5			8	+			6			1	1	1	
Heterolepa pseudoungeriana (CUSHMAN)					+	2		5	2																							
Hoeglundina elegans (D'ORBIGNY)	12	2		+	9	10	1	8	11			7	7		2	2		1	7									1				
Hyalinea balthica (SCHROETER)																																
Lagena s.l.	6	4	6	4	3	5	5	3	2	5	4	7	2	2	6	5	2	3	2	5	4	3	3	3	+	4	3	4	3	10	1	4
Laticarinina pauperata (PARKER & JONES)															1																	
Lenticulina peregrina (SCHWAGER)																																
Lenticulina s.l.	+	2	1	1	1	1	2		1	+			+		+	+	2			1		1	1	+		+	1		+	1	2	+
Lingulina seminuda HANTKEN																																
Melonis affinis (REUSS)		7	9	1		1	14			1	4		+		1	5	5	2	1			4			1	1			2	1		4
Melonis pompilioides (WILLIAMSON)	4	8	18	12	3	3	9	2	5	5	6	5	5	2	+		7	3	4	11	2			8	1				+	7	7	4
Miliolidae		+	2	+	+				2		+		1	1	+		+			1			+								1	
Nodosaria longiscata D'ORBIGNY																																
Nodosaria (incl. Dentalina)		1		+										1		1		+	+			1	+				1	2	+			1
Nummoloculina irregularis (D'ORBIGNY)		+		2	1	+		1	1	9	11	2	3	6			1		1	7				5						1		
Nuttallides umboniferus (CUSHMAN)	1														+	1						5	4			6	17		22	10		1
Ophthalmidium pusillum (EARLAND)																													2			
Oridorsalis umbonatus (REUSS)	11	6	7	8	17	6	6	10	10	7	6	16	22	11	23	25	15	17	11	18	5	20	12	15	8	25	16	18	17	17	7	35
Osangularia culter (PARKER & JONES)																																
Planulina ariminensis D'ORBIGNY																																
Planulina wuellerstorfi (SCHWAGER)	19	14	16	11	9	8	9	7	13	15	16	16	24	24	21	13	7	14	32	20	6	24	11	29	13	13	13	42	6	9	28	10
Pleurostomella alternans SCHWAGER																																
Polymorphinidae																																
Praeglobobulimina pupoides (D'ORBIGNY)		9	6	5	1	1					1			2																		
Pseudonodosaria torrida (CUSHMAN)																																
Pullenia bulloides (D'ORBIGNY)	5	2		1	4	2	2	7	3	3	1	2	2	1	3	2	2	2	2	2		1	5		1	4	2	1	6	7	11	1
Pullenia quinqueloba (REUSS)	4	4	1		4			2	3	1	1		2	1	2	3		2			7	2	10	+	6			1		6	2	2
Pyrgo lucernula (SCHWAGER)								1		+			+																			
Pyrgo murrhyna (SCHWAGER)	2	4	8	6	2	1	5	1	3	11	8	4	2	10	9	8	6	2	4	7	2	5	+	19	2	3	2		6	4		2
Pyrgo serrata (BAILEY)				+		+	+			1																						
Pyrgo spp. indet.	+			1				+				+																				
Pyrgoella sphaera (D'ORBIGNY)																																
Pyrulina cylindroides (ROEMER)						1						1							1										+			
Quinqueloculina spp. indet.	2	2	2	3	3	2		3	1	6	5	9	6	13	+	4	3	+	2		1	+		4	1	+	1	4	3	1	2	
Rotamorphina laevigata (PHLEGER & PARKER)				1					1	+																						
Robertina bradyi CUSHMAN & PARKER																																
Saracenaria italica DEFRANCE																																
Siphonodosaria spp. indet.																																
Siphouvigerina auberiana (D'ORBIGNY)		3		4	1	5	6		2	8			+							13				+							1	3
Sphaeroidina bulloides D'ORBIGNY	+	+	+	1	1		+	2	+			+	+		+		1		1				1	1						1		
Spirillina decorata BRADY																																
Spiroloculina spp. indet.																																
Spirosigmoilina tenuis (CZJZEK)																																
Trifarina angulosa (WILLIAMSON)																																
Triloculina tricarinata D'ORBIGNY																																
Vaginulina spinigera BRADY																																
Valvulineria arctica GREEN																																
Valvulineria complanata (D'ORBIGNY)																																
SHALLOW WATER BENTHONIC FORAMINIFERIDA (%)	0	0	0	0	0	0	0	0	0	0	0	0	0	0	0	0	0	0	0	0	0	0	0	0	0	0	0	0	0	0	0	0
Ammonia beccarii (LINNE)																																
Amphistegina lessonii D'ORBIGNY																																
Asterigerinata mamilla (WILLIAMSON)																																
Cancris oblongus (WILLIAMSON)																																
Cibicides ex gr. lobatulus (WALKER & JACOB)																																
Cribroelphidium gunteri (COLE)																																
Cribrononion advenum (CUSHMAN)																																
Cribrononion incertum (WILLIAMSON)																																
Dyocibicides spec. indet.																																
Elphidiidae																																
Elphidium crispum (LINNE)																																
Eponides repandus (FICHTEL & MOLL)																																
Florilus asterizans (FICHTEL & MOLL)																																
Florilus atlanticus (CUSHMAN)																																
Hanzawaia ex gr. concentrica (CUSHMAN)																																
Miliolidae																																
Neoconorbina orbicularis (TERQUEM)																																
Nonionella spec. indet.																																
Pararotalia armata (D'ORBIGNY)																																
Planorbulina mediterranensis D'ORBIGNY																																
Poritextularia panamensis (CUSHMAN)																																
Ptychomiliola separans (BRADY)																																
Quinqueloculina spp. indet.																																
Reussella miocenica CUSHMAN																																
Rosalina globularis D'ORBIGNY																																
Textulariidae																																
Number of Radiolaria per gramme sediment	22 000	22 000	20 000	40 000	4700	250	7000	30	20	2	960	3000	5400	2600	2300	2000	900	1200	2	230	+	310	+	310	1	0	0	0	29	3900	350	1200
Echinoidea (spines)		33	+	+	+	41	74	+	+	2	1	+	1	23	12	21	3	19	12	1	+	47	4		+	10	2	+	2	3	+	2
Echinoidea (plates)		+																														
Mollusca				16	17	56	2	+	22	11	+	49	540	+	2	+			3													
Ostracoda	+	+	+	+	1	9	2	+	+	1		+	1	+	+	2	+	1	+			1	+	+		+	+			+		
Pisces (teeth)					+		+					+						+							+			+			1	
Pisces (otoliths)	+	+	+	+	+	+		+	+		+	+	+	+		+	+		+		+											
Bryozoa																																

Table 6

Table 7

Samples with a residue of grey shale with predominantly Epistominella exigua

(included in Planulina wuellerstorfi fauna)

Sta	Sample interval	Depth m	W%R	FN	PFN	% P	BFN	% shallow	% Pwu	FGR	FN (fraction > 75 μ)	
2	214-217	4.670	+	1	+		+	-	43	1.400	13	
7	47-50	5.020	+	+	+	100	0	-	x	x	7	
	234-238		+	+	+	100	0	-	x	x	10	
56	30-34	5.020	+	1	+	15	1	-	16	x	9	
62	240-243	4.408	+	+	+	?	+	-	57	x	7	
64	103-106	4.615	+	+	+	0	1	-	32	x	8	(1)

(1) with reworked planktonics.

Samples with a residue of yellow grey shale with Planulina wuellerstorfi fauna

Sta	Sample interval	Depth m	W%R	FN	PFN	% P	BFN	% shallow	% Pwu	FGR	FN (fraction > 75 μ)
58	40-43	4.400	4	150	120	80	32	-	78	4.800	330
64	15-18	4.615	8	950	880	92	73	-	78	14.000	1.400

Sample with a residue of (limonitic?) yellow-grey and brown shale with P. wuellerstorfi fauna

Sta	Sample interval	Depth m	W%R	FN	PFN	% P	BFN	% shallow	% Pwu	FGR	FN (fraction > 75 μ)
5	102-105	3.860	4	650	550	87	90	-	44	19.000	2.700

Samples with a residue of glauconitic shale with Planulina wuellerstorfi fauna

Sta	Sample interval	Depth m	W%R	FN	PFN	% P	BFN	% shallow	% Pwu	FGR	FN (fraction > 75 μ)
5	7-10	3.860	15	530	490	93	36	-	48	7.800	1.400
	44-47		23	3	1	36	2	-	37	16	14
	60-63		2	18	2	11	16	-	26	1.700	200
	70-73		2	390	330	84	63	-	37	28.000	1.400

Samples with a large residue consisting of foraminiferal tests only, with P. wuellerstorfi fauna

Sta	Sample interval	Depth m	W%R	FN	PFN	% P	BFN	% shallow	% Pwu	FGR	FN (fraction > 75 μ)
35	20-28	2.829	67	26.000	26.000	99	260	-	37	50.000	110.000
	90-98		56	30.000	30.000	99	280	-	25	65.000	55.000
	150-158		66	16.000	16.000	99	230	-	26	29.000	36.000
	246-254		56	45.000	45.000	100	170	-	45	100.000	69.000

Samples with a residue of grey shale with Planulina wuellerstorfi fauna(2)

Sta	Sample interval	Depth m	W%R	FN	PFN	% P	BFN	% shallow	% Pwu	FGR	FN (fraction > 75 μ)	
2	8-11	4.670	7	820	800	98	21	-	66	19.000	2.200	*
	37-40		1	16	15	91	2	-	45	5.200	160	
	155-158		4	2.200	2.200	99	18	-	73	230.000	4.900	*
4	12-15	5.070	29	5.700	5.700	99	38	-	53	87.000	9.900	*
	44-47		2	410	400	99	6	-	71	29.000	1.000	
	155-158		5	1.900	1.900	99	11	-	51	45.000	2.900	*
5	80-83	3.860	4	280	250	85	22	+	40	13.000	950	
6	16-19	5.080	6	590	570	97	15	-	63	17.000	1.500	*

(2) in fine fraction often much foraminiferal debris
* residue > 90% foraminiferal tests

Table 8

Samples with a residue of grey shale with Planulina wuellerstorfi fauna (continued) (1)

Sta	Sample interval	Depth m	W%R	FN	PFN	% P	BFN	% shallow	% Pwu	FGR	FN (fraction > 75 μ)	
7	6-10	5.020	11	1.300	1.300	98	32	-	72	16.000	2.400	*
	57-60		9	3.200	3.200	99	23	-	57	47.000	5.700	*
	153-156		3	540	540	99	5	-	53	19.000	1.300	
	210-213		3	830	810	98	20	-	83	37.000	2.000	*
8	20-24	5.084	7	310	270	89	34	-	71	7.000	490	*
	79-82		1	220	220	99	3	-	53	38.000	1.100	*
10	8-11	4.257	9	1.300	1.200	96	52	-	57	23.000	2.700	*
	35-38		1	39	37	95	2	-	45	9.500	360	
	87-91		3	790	780	99	12	-	62	40.000	2.900	
	173-177		1	63	62	98	1	-	48	17.000	290	
11	5-8	3.897	17	1.500	1.400	97	49	-	45	14.000	6.500	*
	71-74		4	230	220	96	8	-	50	28.000	890	*
	127-130		1	120	110	93	8	-	57	25.000	1.200	
	231-234		+	7	6	91	1	-	36	x	30	
12	17-20	1.810	4	800	770	95	40	-	22	29.000	4.300	*
13	25-30	3.200	12	1.900	1.900	98	37	-	35	25.000	14.000	*
	107-111		23	6.000	5.700	96	250	-	23	45.000	12.000	*
	178-182		5	1.600	1.500	97	51	-	27	54.000	17.000	*
17	11-15	3.338	5	1.100	1.100	99	16	-	27	29.000	4.000	
	110-114		2	410	390	97	14	-	35	31.000	630	
	187-191		3	630	620	98	15	-	21	32.000	2.900	
	256-259		2	640	640	99	8	-	35	45.000	3.200	
18	138-144	2.790	1	50	47	96	2	-	38	24.000	490	
	236-242		4	290	250	87	38	-	24	21.000	2.500	
19	10-18	3.626	12	1.500	1.500	99	13	-	48	15.000	7.800	
	80-87		6	1.300	1.300	99	14	-	59	35.000	15.000	*
	150-158		5	890	880	99	9	-	42	25.000	3.500	
20	99-105	2.587	10	900	800	90	120	-	20	18.000	6.100	
	205-211		11	1.300	1.300	97	39	-	30	20.000	7.100	*
21	156-162	2.430	9	1.500	1.500	94	84	-	18	24.000	4.900	*
22	117-123	1.630	+	32	17	47	15	-	14	12.000	100	
23	4-12	1.340	1	110	79	75	27	-	18	24.000	580	

(1) in fine fraction often much foraminiferal debris;
* residue > 90% foraminiferal tests

Table 9

Samples with a residue of grey shale with Planulina wuellerstorfi fauna (continued) (1)

Sta	Sample interval	Depth m	W%R	FN	PFN	% P	BFN	% shallow	% Pwu	FGR	FN (fraction > 75 μ)	
23	100-108	1.340	+	44	34	78	10	-	17	30.000	240	
27	7-15	2.552	8	1.300	1.300	97	39	-	22	19.000	5.500	
29	5-13	3.129	15	1.600	1.500	97	42	-	36	14.000	7.100	
	52-60		13	2.900	2.900	99	31	-	30	24.000	20.000	
32	0-8	3.509	26	2.900	2.900	98	51	-	41	14.000	8.500	*
33	8-16	3.876	15	3.600	3.600	99	42	-	50	33.000	8.100	*
	40-48		15	7.100	7.000	99	40	-	53	62.000	15.000	*
	150-158		4	500	550	98	9	-	32	18.000	2.200	
34	16-24	3.846	22	2.200	2.100	98	39	-	47	13.000	5.800	
	110-118		15	1.800	1.800	99	26	-	58	17.000	7.200	
	200-208		13	2.500	2.400	98	52	-	42	28.000	9.600	
37	3-11	4.430	11	2.700	2.600	99	28	-	61	32.000	6.400	*
	46-54		4	560	540	97	13	-	61	18.000	1.300	
38	6-14	3.252	15	6.000	5.900	99	60	-	45	48.000	16.000	*
	97-105		14	4.400	4.400	98	72	-	34	38.000	16.000	
	170-178		13	5.300	5.300	99	50	-	46	54.000	23.000	*
	240-248		10	2.600	2.600	98	63	-	35	36.000	11.000	
39	3-13	4.617	15	1.500	1.500	98	26	-	62	12.000	2.900	
40	20-26	4.635	11	3.900	3.900	99	42	-	60	45.000	6.000	*
41	23-29	4.658	2	110	110	95	6	-	78	10.000	230	
42	25-32	4.877	17	5.500	5.400	99	74	-	66	41.000	12.000	*
	70-78		10	360	350	98	8	-	45	3.900	510	
	85-90		+	18	17	92	2	-	52	5.700	38	
43	28-34	4.658	5	640	620	97	22	-	65	16.000	1.100	*
	100-106		8	2.000	2.000	98	34	-	71	40.000	5.300	
	186-192		+	29	25	86	4	-	85	24.000	100	
	244-250		15	900	820	91	80	-	77	8.400	3.100	*
44	28-32	4.479	11	390	350	91	37	-	72	4.000	660	
	110-115		1	72	65	91	7	-	70	8.100	340	
45	5-11	3.388	1	87	84	97	3	-	35	8.100	190	
	84-90		2	360	340	95	17	-	39	27.000	1.200	
	155-161		3	420	400	94	23	-	23	20.000	1.700	
	240-246		2	270	200	76	62	-	19	24.000	1.100	
50	8-14	1.973	2	120	100	81	26	-	15	29.000	980	
54	10-15	4.708	16	2.200	2.100	97	60	-	68	16.000	3.500	*
55	31-36	4.728	1	47	38	83	9	-	76	17.000	190	*
	90-95		+	9	8	83	1	-	81	9.400	27	
	153-158		+	11	8	67	4	-	64	11.000	57	
57	10-13	4.896	13	3.800	3.700	98	77	-	58	35.000	6.500	*

(1) in fine fraction often much foraminiferal debris
* residue > 90% foraminiferal tests

Table 10

Samples with a residue of grey shale with Planulina wuellerstorfi fauna (continued) (1)

Sta	Sample interval	Depth m	W%R	FN	PFN	% P	BFN	% shallow	% Pwu	FGR	FN (fraction > 75 μ)	
58	8-11	4.414	4	63	31	49	32	-	68	2.700	210	
62	24-28	4.408	2	36	11	31	25	-	76	4.700	400	*
	86-89		4	320	310	97	10	-	67	16.000	770	
63	1-5	4.639	1	10	8	82	2	-	69	1.900	30	
	50-53		4	170	140	85	26	-	69	6.500	470	*
	134-137		10	330	300	91	29	-	65	6.000	1.100	
	250-253		18	860	810	95	42	-	49	7.200	2.400	
64	179-182	4.615	1	7	6	90	1	-	45	x	30	
65	7-10	3.838	4	330	320	98	6	-	27	16.000	1.600	
	91-94		6	740	710	96	30	-	35	35.000	4.200	
	235-238		+	68	52	77	16	-	26	26.000	170	
66	20-23	3.442	2	380	360	96	16	-	56	22.000	1.000	*
	107-110		3	360	140	89	18	-	55	10.000	710	
	177-180		2	58	35	60	23	-	17	8.300	200	
	250-253		+	16	10	63	6	-	99	7.600	40	

(1) in fine fraction often much foraminiferal debris
* residue > 90% foraminiferal tests

Table 11

Samples with a residue consisting mainly of sand

Sta	Sample interval	Depth m	W%R	FN	PFN	% P	BFN	% shallow	% Pwu	FGR	FN (fraction >75 µ)
27	100-104	2.552	98	40	27	69	12	66	5	47	120
28	35-43	3.240	35	810	780	97	26	2	26	2.800	12.000
	67-71		91	1.200	1.100	90	120	36	8	2.300	2.400
29	103-197	3.129	92	80	49	63	31	68	1	120	570
	212-220		39	780	770	98	18	12	20	3.100	3.600
32	36-40	3.509	89	580	510	89	67	27	7	960	2.000
36	35-39	4.540	81	410	360	86	48	41	5	740	1.000
37	74-78	4.430	66	450	360	79	96	19	11	21.000	3.100
39	31-38	4.617	93	180	150	82	32	45	4	260	410
41	55-61	4.658	96	65	29	45	36	61	2	70	170
	140-146		96	110	63	58	43	48	3	120	280
52	7-10	4.316	88	210	35	16	180	68	1	400	2.500
	65-70		97	140	41	29	100	72	1	160	630
	107-110		95	130	35	26	98	71	0	170	850
	144-148		96	140	47	34	90	71	0	170	640
53	4-7	4.602	88	590	220	38	360	60	0	3.200	3.600
	63-66		16	130	88	69	40	36	11	1.300	460
56	102-105	5.020	92	250	36	14	220	80	0	390	1.700
	173-176		92	340	85	23	260	85	0	520	1.600
	210-213		83	1.200	490	40	740	63	1	5.700	5.600
57	50-53	4.896	86	430	160	37	270	78	0	670	2.600
	84-87		8	5	1	21	4	81	0	410	92
	120-123		84	810	230	28	580	72	1	1.600	4.300
	187-190		91	350	81	31	250	75	0	480	2.200
59	5-8	4.452	92	200	56	30	140	71	1	310	1.400

Samples with a sandy residue

Sta	Sample interval	Depth m	W%R	FN	PFN	% P	BFN	% shallow	% Pwu	FGR	FN (fraction >75 µ)
27	62-70	2.552	39	580	560	97	19	8	18	5.400	2.000
32	100-106	3.509	24	200	180	90	20	2	13	14.000	710
36	10-18	4.540	39	6.300	6.300	99	56	2	60	17.000	13.000
39	72-80	4.617	2	99	97	98	2	4	80	11.000	550
	132-142		9	23	21	91	2	4	7	400	230
40	110-116	4.635	7	200	180	87	27	1	65	8.500	680
44	180-186	4.479	+	15	12	80	3	-	80	8.600	41
53	20-25	4.602	6	78	67	85	11	5	59	1.600	230
54	80-85	4.708	2	50	47	95	3	6	72	5.500	140
55	234-239	4.728	1	50	39	78	11	+	66	12.000	220

Table 12

Samples with a sandy residue (continued)

Sta	Sample interval	Depth m	W%R	FN	PFN	% P	BFN	% shallow	% Pwu	FGR	FN (fraction > 75µ)
56	55-60	5.020	+	+	+	38	+	-	28	x	3
	89-91		4	71	58	83	13	65	12	7.500	240
	131-134		2	180	180	98	4	1	62	23.000	390
	189-192		1	49	40	81	10	2	89	14.000	310
57	207-210	4.896	2	21	15	72	6	76	9	1.700	120
58	61-64	4.414	25	120	74	64	42	31	6	800	650
62	148-151	4.408	41	1.600	700	44	900	4	3	31.000	28.000
63	195-198	4.639	3	12	7	63	5	3	60	840	86

Samples with a residue of plant material

Sta	Sample interval	Depth m	W%R	FN	PFN	% P	BFN	% shallow	% Pwu	FGR	FN (fraction > 75µ)
14	16-19	1.497	5	150	136	87	20	-	16	6.000	810
	102-105		4	92	78	85	14	-	17	6.600	350
15	25-29	1.441	1	130	110	83	23	-	18	29.000	590
	164-167		1	110	94	84	18	-	7	17.000	1.100
	253-256		5	54	45	83	9	6	5	19.000	440
42	121-124	4.877	10	5	3	70	2	2	5	550	1.500

Samples with a residue of almost barren shale

Sta	Sample interval	Depth m	W%R	FN	PFN	% P	BFN	% shallow	% Pwu	FGR	FN (fraction > 75µ)
6	50-53	5.080	+	1	0	0	1	72	16	x	2
64	248-252	4.615	+	0	0	-	0	25	0	x	+

Samples with a residue of grey shale-rich in Mollusca, etc.

Sta	Sample interval	Depth m	W%R	FN	PFN	% P	BFN	% shallow	% Pwu	FGR	FN (fraction > 75µ)
65	156-158	3.838	67	12.000	7.700	66	3.900	7	4	21.000	15.000

Core Adriatic Sea (BROUWER, 1967)

Sta	Sample interval	Depth m	W%R	FN	PFN	% P	BFN	% shallow	% Pwu	FGR	FN (fraction > 75µ)
293	II	1198	36	500	260	50	240	37	-	680	10.000
	III		30	1.000	450	45	550	34	-	3.400	6.100

Remark: Samples with 15% Pwu are also considered as containing the *Planulina wuellerstorfi* fauna.

Table 13

Samples with a residue of grey shale with faunas of group XI (1)

Sta	Sample interval	Depth m	W%R	FN	PFN	% P	BFN	% shallow	% Pwu	FGR	FN (fraction > 75μ)	
1	12-15	2.966	1	150	140	94	9	-	18	38.000	620	*
	114-117		2	160	100	66	52	-	21	18.000	780	*
12	100-103	1.810	11	2.700	2.500	93	170	-	14	32.000	11.000	*
	143-146		12	3.200	2.900	93	220	-	11	43.000	32.000	*
	230-233		3	430	350	82	78	-	11	30.000	2.700	
14	172-175	1.497	9	320	270	85	49	-	10	45.000	890	
15	108-111	1.441	2	440	380	86	61	-	13	43.000	1.900	*
18	45-51	2.790	19	4.100	3.900	96	160	-	15	33.000	15.000	*
20	40-46	2.597	20	3.300	3.300	98	67	-	14	22.000	22.000	*
21	3-9	2.430	17	1.600	1.500	95	75	-	18	14.000	4.000	
	59-65		34	5.600	5.300	96	220	-	19	20.000	40.000	*
	250-256		4	820	770	92	59	-	9	40.000	4.700	
22	10-16	1.630	2	240	210	88	28	-	13	26.000	1.300	
	200-206		+	7	2	34	4	-	10	19.000	28	
23	170-178	1.340	+	74	55	74	19	-	13	41.000	350	
24	0-8	692	1	140	51	36	91	-	2	18.000	700	
25	5-15	680	1	81	36	45	44	-	3	21.000	370	
	74-82		1	100	49	49	53	-	4	22.000	410	
	170-178		+	63	19	30	44	-	-	21.000	320	
26	4-14	337	2	180	40	23	140	5	1	20.000	1.600	
46	10-15	918	6	340	54	16	280	-	2	5.900	1.000	
	117-127		18	3.000	2.100	69	940	-	5	24.000	23.000	
	186-196		18	5.400	4.400	79	920	1	2	42.000	52.000	
	270-280		12	2.000	1.700	82	370	-	5	21.000	7.300	
49	5-11	1.502	1	49	5	10	44	-	4	13.000	220	
	113-118		7	2.200	2.000	88	270	-	15	42.000	18.000	
	202-208		6	2.200	1.900	86	300	-	7	53.000	11.000	
50	70-75	1.973	1	280	230	84	44	-	11	28.000	1.700	
	188-194		6	920	750	80	180	-	4	25.000	7.000	
51	8-13	2.492	2	290	270	93	19	-	15	30.000	1.900	
	111-116		5	700	630	91	69	-	7	21.000	2.600	
	150-155		2	490	400	82	87	-	16	31.000	2.000	
	256-261		2	340	210	62	130	-	18	32.000	630	

Samples with a residue of grey shale, rich in Mollusca, etc.

Sta	Sample interval	Depth m	W%R	FN	PFN	% P	BFN	% shallow	% Pwu	FGR	FN (fraction > 75μ)
24	74-82	692	1	200	130	63	74	-	5	27.000	1.200
	171-179		9	1.300	1.000	79	270	4	7	28.000	12.000
26	95-103	337	2	390	170	45	210	5	2	37.000	3.000

(1) in fine fraction much foraminiferal debris.

* residue > 90% foraminiferal tests

Table 14

Samples with a residue rich in pellets and Mollusca, etc.

Sta	Sample interval	Depth m	W%R	FN	PFN	% P	BFN	% shallow	% Pwu	FGR	FN (fraction > 75 μ)
47	10-15	304	74	17.000	6.200	36	11.000	8	6	26.000	35.000
	99-105		76	12.000	4.900	41	7.000	9	6	18.000	26.000
	165-171		26	5.000	2.300	46	2.700	2	6	22.000	9.100

Samples with a residue rich in pellets

Sta	Sample interval	Depth m	W%R	FN	PFN	% P	BFN	% shallow	% Pwu	FGR	FN (fraction > 75 μ)
48	3-9	548	25	6.200	510	8	5.700	1	-	34.000	30.000
	80-86		12	1.800	160	9	1.700	1	1	20.000	12.000
	145-151		9	2.000	290	15	1.700	+	1	30.000	5.000

Core Adriatic Sea (BROUWER, 1967)

Sta	Sample interval	Depth m	W%R	FN	PFN	% P	BFN	% shallow	% Pwu	FGR	FN (fraction > 75 μ)
293	I	1198	+	130	120	93	9	1	-	>18.000	860

APPENDIX

CHECKLIST OF FORAMINIFERA

In the following list, only one or two references are made to illustrations in the literature in an attempt to restrict ourselves to two publications. Brady's work on the Recent Foraminifera collected during the Challenger expedition still gives the best illustrations, although the nomenclature has changed since considerably. Therefore, as a second reference we chose Phleger, Parker & Peirson's study on Atlantic Foraminifera. The majority of the species found in the Gulf of Guinea cores are illustrated in these two publications. We also made extensive use of R.W. Barker's Taxonomic Notes on the species figured by Brady, although this is not shown in the following literature list.

i. Planktonic species

Candeina nitida d'Orbigny
Candeina nitida d'Orbigny, Brady, 1884, pl. LXXXII, figs. 13-20;
Phleger, Parker & Peirson, 1953, pl. 2, figs. 22, 23.

Globigerina bulloides d'Orbigny
Globigerina bulloides d'Orbigny, Brady, 1884, pl. LXXIX, figs. 3-7; Phleger, Parker & Peirson, 1953, pl. 1, figs. 3, 4, 7, 8.

Globigerina inflata d'Orbigny
Globigerina inflata d'Orbigny, Brady, 1884, pl. LXXIX, figs. 8-10; Phleger, Parker & Peirson, 1953, pl. 1, figs. 15, 16.

Globigerina pachyderma (Ehrenberg)
Globigerina pachydermå (Ehrenberg), Brady, 1884, pl. CXIV, figs. 19, 20;
Globigerina pachyderma (Ehrenberg), Phleger, Parker & Peirson, 1953, pl. 1, figs. 17-19.

Globigerinella aequilateralis (Brady)
Globigerina aequilaralis Brady, Brady, 1884, pl. LXXX, figs. 18-21;
Globigerinella aequilateralis (H.B. Brady), Phleger, Parker & Peirson, 1953, pl. 2, figs. 9-11.

Globigerinita glutinata (Egger)
Globigerinita glutinata (Egger), Phleger, Parker & Peirson, 1953, pl. 2, figs. 12-15.

Globigerinoides conglobatus(Brady)
Globigerina conglobata Brady Brady, 1884, pl. LXXXX, figs. 1-5;
Globigerinoides conglobata (H.B. Brady), Phleger, Parker & Peirson, 1953, pl. 2, figs. 1-3.

Globigerinoides extremus Bolli & Bermudez
Globigerinoides obliquus extremus Bolli & Bermudez, 1965, pl. 1, figs. 10-12.

Globigerinoides immaturus Leroy
Globigerinoides sacculiferus (Brady) var. immatura Leroy, 1939, pl. 3, figs. 19-21.

Globigerinoides ruber (d'Orbigny)
Globigerina rubra d'Orbigny, Brady, 1884, pl. LXXIX, figs. 11-16;
Globigerinoides rubra (d'Orbigny), Phleger, Parker & Peirson, 1953, pl. 2, figs. 4, 7.
Only pink-coloured specimens are comprised in this species.

Globigerinoides sacculifer (Brady)
Globigerina sacculifera Brady, Brady, 1884, pl. LXXX, figs. 11-17;
Globigerinoides sacculifera (H.B. Brady), Phle,ger, Parker & Peirson, 1953, pl. 2, figs. 5, 6.

Globigerinoides trilobus (Reuss)
Globigerina bulloides d'Orbigny var. triloba Reuss, Brady, 1884, pl. XXIX, figs. 1, 2.
This species also includes non-coloured specimens of G. ruber.

Globoquadrina altispira (Cushman & Jarvis)
Globigerina altispira Cushman & Jarvis, Phleger, Parker & Peirson, 1953, pl. 1, figs. 1, 2, 6.

Globoquadrina dutertrei (d'Orbigny)
Globigerina dubia Egger, Brady, 1884, pl. LXXIX, fig. 17.
Globigerina eggeri Rhumbler, Phleger, Parker & Peirson, 1953, pl. 1, figs. 11, 12.

Globorotalia acostaensis Blow
Globorotalia acostaensis Blow, 1959, pl. 17, figs. 106a-c.

Globorotalia margaritae Bolli & Bermudez
Globorotalia margaritae Bolli & Bermudez, 1965, pl. 1, figs. 16-18.
Some specimens have been found in core 5, 102-105 cm. The species is not mentioned on the distribution chart.

Globorotalia menardii (d'Orbigny)
Pulvinulina menardii (d'Orbigny), Brady, 1884, pl. CIII, figs. 1, 2.
Globorotalia menardii (d'Orbigny), Phleger, Parker & Peirson, 1953, pl. 3, figs. 1, 2, 4, 5.

Globorotalia punctulata (d'Orbigny)
Pulvinulina crassa (d'Orbigny), Brady, 1884, pl. CIII, figs. 11, 12.
Globorotalia punctulata (d'Orbigny), Phleger, Parker & Peirson, 1953, pl. 4, figs. 8-12.

Globorotalia scitula (Brady)
Pulvinulina patagonica (d'Orbigny), Brady, 1884, pl. CIII, fig. 7.
Globorotalia scitula (H.B. Brady), Phleger, Parker & Peirson, 1953, pl. 4, figs, 13, 14.

Groborotalia truncatulinoides (d Orbigny)
Pulvinulina micheliana (d'Orbigny), Brady, 1884, pl. CIV, figs. 1,2.
Globorotalia truncatulinoides (d'Orbigny), Phleger, Parker & Peirson, 1952, pl. 4, figs. 17, 18.

Globorotalia tumida (Brady)
Pulvinulina tumida Brady, Brady, 1884, pl. CII, figs. 4-6.
Globorotalia tumida (H.B. Brady), Phleger, Parker & Peirson, 1953, pl. 3, figs. 3, 6, 7, 8, 10, 11.
This species, which does not appear on the distribution charts is included in *G. menardii.* It occurs abundantly in the "warm" intervals of the Pleistocene.

Orbulina bilobata (d'Orbigny)
Orbulina bilobata (d'Orbigny), Bolli, 1957, pl. 27, fig. 6.

Orbulina universa d'Orbigny
Orbulina universa d'Orbigny, Brady, 1884, pl. LXXXI, figs. 8-19, 22-26, pl. LXXXII, fig. 1; Phleger Parker & Peirson, 1953, pl. 2, fig. 8.

Pulleniatina obliquiloculata (Parker & Jones)
Pullenia obliquiloculata Parker & Jones, Brady, 1884, pl. LXXXIV, figs. 16-20;
Pulleniatina obliquiloculata (Parker & Jones), Phleger, Parker & Peirson, 1953, pl. 2, figs. 16-18.

Pulleniatina primalis Banner & Blow
Pulleniatina primalis Banner & Blow, 1967, pl. 1, figs. 3-8, pl. 3, fig. 2.

Sphaeroidinella dehiscens (Parker & Jones)
Sphaeroidina dehiscens Parker & Jones, Brady, 1884, pl. LXXXIV, figs. 8-11.
Sphaeroidinella dehiscens (Parker & Jones), Phleger, Parker & Peirson, 1953, pl. 2, fig. 19.

Sphaeroidinella subdehiscens Blow
Sphaeroidinella dehiscens (Parker & Jones) subdehiscens Blow, 1959, pl. 12, figs, 71a-c.

Tretomphalus atlanticus Cushman
Tretomphalus atlanticus Cushman, Phleger, Parker & Peirson, 1953, pl. 9, figs. 30, 31.
This species, which has been found for example in core 62, 148-151 cm, does not appear on the distribution charts.

ii. *Agglutinated benthonic species*

Adercotryma glomerata (B r a d y)
Haplophragmium glomeratum Brady, B r a d y, 1884, pl. XXXIV, figs. 15-18.

Alveolophragmium scitulum (B r a d y)
Haplophragmium scitulum Brady, B r a d y, 1884, pl. XXXIV, figs. 11-13.
Occurs in core 32, 0-8 cm; not shown on distribution charts.

Ammomarginulina foliacea (B r a d y)
Haplophragmium foliaceum Brady, B r a d y, 1884, pl. XXXIII, figs. 20-25.
Found in core 11, 5-8 cm; not mentioned on distribution charts.

Crisbrostomoides subglobosum (S a r s)
Haplophragmium latidorsatum (Bornemann), B r a d y, 1884, pl. XXXIV, figs. 7-10.

Cyclammina trullisata (B r a d y)
Trochammina trullisata Brady, B r a d y, 1884 pl. XL, figs. 13-16.

Cystammina galeata (B r a d y)
Trochammina galeata Brady, B r a d y, 1884, pl. XL, figs. 19-23.

Eggerella bradyi (C u s h m a n)
Verneuilina pygmaea (Egger), B r a d y, 1884, pl. XLVII, figs. 4-7;
Eggerella bradyi (Cushman), P h l e g e r, P a r k e r & P e i r s o n, 1953, pl. 5, figs. 8, 9.
Specimens of Karreriella bradyi (C u s h m a n) have been included in this species.

Gaudryina atlantica (B a i l e y)
Verneuilina triquetra (Münster), B r a d y, 1884, pl. XLVII, fig. 18.

Glomospira charoides (J o n e s & P a r k e r)
Ammodiscus charoides (Jones & Parker), B r a d y, 1884, pl. XXXVIII, figs. 10-16;
Glomospira charoides (Jones & Parker), P h l e g e r, P a r k e r & P e i r s o n, 1953, pl. 5, fig. 1.

Hyperammina elongata B r a d y
Hyperammina elongata Brady, B r a d y, 1884, pl. XXIII, figs. 4, 7-10.

Karreriella apicularis (C u s h m a n)
Gaudryina siphonella Reuss, B r a d y, 1884, pl. XLVI, figs. 17-19;
Plectina apicularis (Cushman), P h l e g e r, P a r k e r & P e i r s o n, 1953, pl. 5, fig. 10.

Martinottiella communis (d'O r b i g n y)
Clavulina communis d'Orbigny, B r a d y, 1884, pl. XLVIII, figs. 3, 4, 6, 7, 8-13.

Recurvoides turbinatus (B r a d y)
Haplophragmium turbinatum Brady, B r a d y, 1884, pl. XXXV, fig. 9.

Rhabdammina abyssorum S a r s
Rhabdammina abyssorum M. Sars, B r a d y, 1884, pl. XXXI, figs. 1-13.

Saccammina sphaerica B r a d y
Saccammina sphaerica M. Sars, B r a d y, 1884, pl. XVIII, figs. 11-17.

Sigmoilopsis schlumbergeri (S i l v e s t r i)
Planispirina celata (Costa), B r a d y, 1884, pl. VIII, figs. 1-4;
Sigmoilina schlumbergeri A. Silvestri, P h l e g e r, P a r k e r & P e i r s o n, 1953, pl. 5, fig. 17.

Siphotextularia rolshauseni P h l e g e r & P a r k e r
Siphotextularia rolshauseni Phleger & Parker, P h l e g e r, P a r k e r & P e i r s o n, 1953, pl. 5, fig. 7.

Trochammina globigeriniformis P a r k e r & J o n e s
Haplophragmium globigeriniforme (Parker & Jones), B r a d y, 1884, pl. XXXV, figs. 10, 11.

iii. *Calcareous benthonic species*

Amphicoryna scalaris (B a t s c h)
Nodosaria scalaris (Batsch), B r a d y, 1884, pl. LXIII, figs. 28-31.

Anomalina globulosa C h a p m a n & P a r r
Anomalina grosserugosa (Gümbel), B r a d y, 1884, pl. XCIV, figs. 4, 5.

Anomalinoides io (C u s h m a n)
Cibicides pseudoungerianus var.io, C u s h m a n, 1931, pl. XXIII, figs. 1, 2.

Astrononion stelligerum (d'O r b i g n y)
Nonionina stelligera d'Orbigny, B r a d y, 1884, pl. CIX, figs. 3-5.

Bolivina albatrossi C u s h m a n
Bolivina albatrossi C u s h m a n, 1922, pl. VI, fig. 4.

Bolivina goësi C u s h m a n
Bolivina goessii C u s h m a n, 1922, pl. VI, fig. 5.

Bolivina pseudoplicata H e r o n-A l l e n & E a r l a n d
Bolivina pseudoplicata Heron-Allen & Earland, P h l e g e r, P a r k e r & P e i r s o n, 1953, pl. 7, figs. 18, 19.

Bolivinita quadrilatera (S c h w a g e r)
Textularia quadrilatera Schwager, B r a d y, 1884, pl. XLII, figs. 8-12.

Brizalina alata (S e g u e n z a)
Bolivina beyrichi var.alata Seguenza, B r a d y, 1884, pl. LIII, figs. 2-4.

Brizalina pettersoni (P a r k e r)
Bolivina pettersoni P a r k e r, in P h l e g e r, P a r k e r & P e i r s o n, 1953, pl. 7, figs. 15-17.

Brizalina subaenariensis (C u s h m a n)
Bolivina aenariensis (Costa), B r a d y, 1884, pl. LIII, figs. 10, 11;
Bolivina subaenariensis Cushman, P h l e g e r, P a r k e r & P e i r s o n, 1953, pl. 7, figs. 24, 25.

Bulimina aculeata d'O r b i g n y
Bulimina aculeata d'Orbigny, B r a d y, 1884, pl. LI, figs. 7-9; P h l e g e r, P a r k e r & P e i r s o n, 1953, pl. 6, figs, 21, 22.

Bulimina alazanensis C u s h m a n
Bulimina buchiana d'Orbigny, B r a d y, 1884, pl. LI. figs. 18, 19;
Bulimina alazanensis Cushman, P h l e g e r, P a r k e r & P e i r s o n, 1953, pl. 6, fig. 23.

Bulimina exilis B r a d y
Bulimina elegans var. exilis B r a d y, 1884, pl. L, figs. 5, 6.
Bulimina exilis H.B. Brady, P h l e g e r, P a r k e r & P e i r s o n, 1953. pl. 6, fig. 24.

Bulimina inflata S e g u e n z a
Bulimina inflata Seguenza, B r a d y, 1884, pl. LI, figs. 10-13;
Bulimina mexicana Cushman, P h l e g e r, P a r k e r & P e i r s o n, 1953. pl. 6, fig. 27.

Bulimina marginata d'Orbigny
Bulimina marginata d'Orbigny, B r a d y, 1884, pl. LI. figs. 3-5;
Bulimina marginata d'Orbigny variant, P h l e g e r, P a r k e r & P e i r s o n, 1953, pl. 6, figs. 25, 26.

Cassidella bradyi (C u s h m a n)
Virgulina subsquamosa Egger, B r a d y, 1884, pl. LII, fig. 9.
Virgulina bradyi Cushman, P h l e g e r, P a r k e r & P e i r s o n, 1953, pl. 7, figs. 4,5.

Cassidella mexicana (C u s h m a n)
Virgulina subsquamosa Egger, B r a d y, 1884, pl. LII, figs. 10, 11;
Virgulina mexicana Cushman, P h l e g e r, P a r k e r & P i e r s o n, 1953, pl. 7, figs. 6, 7.

Cassidulina carinata S i l v e s t r i
Cassidulina laevigata d'Orbigny, B r a d y, 1884, pl. LIV, figs. 2, 3.
Cassidulina carinata A. Silvestri, P h l e g e r, P a r k e r &

P e i r s o n, 1953, pl. 9, figs. 32, 37.

Cassidulina curvata P h l e g e r & P a r k e r

Cassidulina curvata Phleger & Parker, P h l e g e r, P a r k e r & P e i r s o n, 1953, pl. 10, fig. 2.

Found in core 35, 150-158 cm, not mentioned on distribution charts.

Cassidulinoides bradyi (N o r m a n)

Cassidulina bradyi Norman, B r a d y, 1884, pl. LIV, figs. 6-9.

Cassidulinoides mexicana (C u s h m a n)

Cassidulinoides mexicana (Cushman), P h l e g e r, P a r k e r & P e i r s o n, 1953, pl. 10, figs. 10, 11.

Ceratobulimina pacifica C u s h m a n & H a r r i s

Bulimina contraria (Reuss), B r a d y, 1884, pl. LIV, fig. 18.

Chilostomella oolina S c h w a g e r

Chilostomella ovoidea Reus, B r a d y, 1884, Pl. LV, figs. 12-14, 17, 18.

Chilostomella oolina Schwager, P h l e g e r, P a r k e r & P e i r s o n, 1953, pl. 10, fig. 18.

Cibicides ex gr. lobatulus (W a l k e r & J a c o b)

Truncatulina lobatula (Walker & Jacob), B r a d y, 1884, pl. XCII, fig. 10, Pl. XCIII, figs. 9. 14.

Cibicides cf. refulgens Montfort, P h l e g e r, P a r k e r & P e i r s o n, 1953, pl. 11, figs. 10, 11.

Some of the material, which is not obviously displaced from shallow water, has been put under this heading.

Cibicidoides robertsonianus (B r a d y)

Truncatulina robertsoniana Brady, B r a d y, 1884, pl. XCV, fig. 4.

Cibicides robertsonianus (H.B. Brady), P h l e g e r, P a r k e r & P e i r s o n, 1953, pl. 11, figs. 15, 16, 17.

Cornuspira carinata (C o s t a)

Cornuspira carinata (Costa), B r a d y, 1884, pl. XI, fig. 4.

Present in core 24, 171-179 cm; not recorded on distribution charts.

Ehrenbergina trigona G o e s

Ehrenbergina serrata Reuss, B r a d y, 1884, pl. LV, figs. 2, 3.

Ehrenbergina trigona Goës, P h l e g e r, P a r k e r & P e i r s o n, 1953, pl. 10, figs. 12-13.

Ehrenbergina undulata P a r k e r

Ehrenbergina undulata P a r k e r, in P h l e g e r, P a r k e r & P e i r s o n, 1953, pl. 10, figs. 14-16.

Epistominella exigua (B r a d y)

Pulvinulina exigua B r a d y, 1884, pl. CIII, figs. 13, 14.

Epistominella exigua (H.B. Brady), Phleger, P a r k e r & P e i r s o n, 1953, pl. 9, figs. 35, 36.

Epistominella smithi (S t e w a r t & S t e w a r t)

Epistominella smithi (R.E. & K.C. Stewart), L i p p s, 1965, pl. II, figs. 6a-c.

Eponides polius P h l e g e r & P a r k e r

Eponides polius Phleger & Parker, P h l e g e r, P a r k e r & P e i r s o n, 1953, pl. 9, figs. 3, 4.

Eponides pusillus P a r r

Eponides pusillus Parr, P h l e g e r, P a r k e r & P e i r s o n, 1953, pl. 9, figs. 5, 6.

Eponides regularis P h l e g e r & P a r k e r

Eponides regularis P h l e g e r & P a r k e r, 1951, pl. 11, figs. 3-4.

Euuvigerina peregrina (C u s h m a n)

Uvigerina pygmaea d'Orbigny, B r a d y, 1884, pl. LXXIV, figs. 11, 12.

Uvigerina hollicki Thalmann and variant, P h l e g e r, P a r k e r & P e i r s o n, 1953, pl. 8, figs. 1-3.

Francesita advena (C u s h m a n)

Virgulina advena Cushman, P h l e g e r, P a r k e r & P e i r s o n, 1953, pl. 7, figs. 1, 2.

Fursenkoina complanata (E g g e r)

Virgulina complanata Egger, P h l e g e r, P a r k e r & P e i r s o n, 1953, pl. 7, fig. 3.

Gavelinopsis translucens (P h l e g e r & P a r k e r)

"Rotalia" translucens Phleger & Parker, P h l e g e r, P a r k e r & P e i r s o n, 1953, pl. 9, figs. 22, 23.

Globobulimina pacifica C u s h m a n

Bulimina pyrula d'Orbigny, B r a d y, 1884, pl. L, figs. 7-10.

Globocassidulina crassa (d'O r b i g n y)

Cassidulina crassa d'Orbigny, B r a d y, 1884, pl. LIV, figs. 4, 5; P h l e g e r, P a r k e r & P e i r s o n, 1953, pl. 10, fig. 1. (mentioned on the distribution charts as Globocassidulina oblonga (Reuss), which is a synonym).

Globocassidulina subglobosa (B r a d y)

Cassidulina subglobosa Brady, B r a d y, 1884, pl. LIV, fig. 17; P h l e g e r, P a r k e r & P e i r s o n, 1953, pl. 10, fig. 4.

Gyroidina lamarckiana (d'O r b i g n y)

Gyroidina lamarckiana (d'Orbigny), P h l e g e r, P a r k e r & P e i r s o n, 1953, pl. 8, figs. 33, 34).

Gyroidina soldanii d'O r b i g n y

Rotalia broeckhiana Karrer, B r a d y, 1884, pl. CVII, fig. 4;

Rotalia soldanii d'Orbigny, B r a d y, 1884, pl. CVII, figs. 6, 7;

Gyroidina soldanii d'Orbigny variants, P h l e g e r, P a r k e r & P e i r s o n, 1953, pl. 9, figs. 1, 2.

Heterolepa pseudoungeriana (C u s h m a n)

Truncatulina ungeriana (d'Orbigny), B r a d y, 1884, pl. XCIV, fig. 9.

Hoeglundina elegans (d'O r b i g n y)

Pulvinulina elegans (d'Orbigny), B r a d y, 1884, pl. CV, figs. 3-6;

Höglundina elegans (d'Orbigny), P h l e g e r, P a r k e r & P e i r s o n, 1953, pl. 9, figs. 24, 25.

Hyalinea balthica (S c h o e t e r)

Operculina ammonoides (Gronovius), B r a d y, 1884, pl. CXII, figs. 1, 2.

Anomalina balthica (Schroeter), P h l e g e r, P a r k e r & P e i r s o n, 1953, pl. 10, figs. 24, 25.

Laticarinina pauperata (P a r k e r & J o n e s)

Pulvinulina pauperata Parker & Jones, B r a d y, 1884, pl. CIV, figs. 3-11;

Laticarinina pauperata (Parker & Jones), P h l e g e r, P a r k e r & P e i r s o n, 1953, pl. 11, figs. 5, 6.

Lenticulina peregrina (S c h w a g e r)

Cristellaria variabilis Reuss, B r a d y, 1884, pl. LXVIII, figs. 11-16.

Lingulina seminuda H a n t k e n

Lingulina carinata var. seminuda Hantken, B r a d y, 1884, pl. LXV, figs. 14, 15.

Melonis affinis (R e u s s)

Nonionina umbilicatula (Montagu), B r a d y, 1884, pl. CIX, figs. 8, 9.

Nonion barleeanum (Williamson), P h l e g e r, P a r k e r & P e,i r s o n, 1953, pl. 6, fig. 4.

Melonis pompilioides (W i l l i a m s o n)

Nonionina pompilioides (Fichtel & Moll), B r a d y, 1884, pl. CIX, figs. 10, 11.

Nonion pompilioides (Fichtel & Moll), P h l e g e r, P a r k e r & P e i r s o n, 1953, pl. 6, figs. 7, 8.

Nodosaria longiscata d'O r b i g n y

Nodosaria longiscata d'O r b i g n y, 1846, pl. i, figs. 10-12;

Lagena elongata (Ehrenberg), B r a d y, 1884, pl. LVI. fig. 29.

Nummoloculina irregularis (d'O r b i g n y)

Biloculina irregularis d'Orbigny, B r a d y, 1884, pl. I, figs. 17, 18.

Nummoloculina irregularis (d'Orbigny), P h l e g e r, P a r k e r & P e i r s o n, 1953, pl. 5, figs. 19, 20.

Nuttallides umboniferus (C u s h m a n)

Truncatulina pygmaea Hantken, B r a d y, 1884, pl. XCV, figs. 9, 10.

Epistominella (?) umbonifera (Cushman), P h l e g e r, P a r k e r & P e i r s o n, 1953, pl. 9, figs. 33, 34.

Ophthalmidium pusillum (E a r l a n d)
Spiroloculina tenuis (Czjzek), B r a d y, 1884, pl. X, fig. 10.
Oridorsalis umbonatus (R e u s s)
Pulvinulina umbonata Reus, B r a d y, 1884, pl. CV, fig. 2;
Eponides umbonatus (Reuss) and variants, P h l e g e r, P a r k e r & P e i r s o n, 1953, pl. 9, figs. 9, 10.
Osangularia culter (P a r k e r & J o n e s)
Truncatulina culter (Parker & Jones), B r a d y, 1884, pl. XCVI, fig. 3;
Osangularia culter (Parker & Jones), P h l e g e r, P a r k e r & P e i r s o n, 1953, pl. 9, figs. 11, 16.
Planulina ariminensis d'O r b i g n y
Anomalina ariminensis (d'O r b i g n y), B r a d y, 1884, pl. XCIII, figs. 10, 11;
Planulina ariminensis d'Orbigny, P h l e g e r, P a r k e r & P e i r s o n, 1953. pl. 11, figs. 3, 4.
Planulina wuellerstorfi (S c h w a g e r)
Truncatulina wuellerstorfi (Schwager), B r a d y, 1884, pl. XCIII, figs. 8, 9;
Planulina wuellerstorfi (Schwager), P h l e g e r, P a r k e r & P e i r s o n, 1953, pl. 11, figs. 1, 2.
Pleurostomella alternans S c h w a g e r
Pleurostomella alternans Schwager, Brady, 1884, pl. LI, fig. 23.
Praeglobobulimina pupoides (d'O r b i g n y)
Bulimina pupoides d'Orbigny, B r a d y, 1884, pl. L, fig. 15.
Pseudonosaria torrida (C u s h m a n)
Nodosaria (Glandulina) laevigata d'Orbigny, 1884, pl. LXI, figs. 20-22.
Pullenia bulloides (d'O r b i g n y)
Pullenia sphaeroides (d'Orbigny), B r a d y, 1884, pl. LXXXIV, figs. 12, 13;
Pullenia bulloides (d'Orbigny), P h l e g e r, P a r k e r & P e i r s o n, 1953, pl. 10, fig. 19.
Pullenia quinqueloba (R e u s s)
Pullenia quinqueloba Reuss, B r a d y, 1884, pl. LXXXIV, figs. 14, 15;
Pullenia quinqueloba (Reuss), P h l e g e r, P a r k e r & P e i r s o n, 1953, pl. 10, fig. 20.
Pyrgo lucernula (S c h w a g e r)
Biloculina bulloides d'Orbigny, B r a d y, 1884, pl. II, figs. 5, 6.
Pyrgo murrhyna (S c h w a g e r)
Biloculina depressa d'Orbigny var. murrhyna Schwager, B r a d y, 1884, pl. II, figs. 10, 11;
Biloculina depressa d'Orbigny, B r a d y, 1884, pl. II, fig. 15.
Pyrgo murrhyna (Schwager), P h l e g e r, P a r k e r & P e i r s o n, 1953, pl. 5, figs. 22-24.
Pyrgo serrata (B a i l e y)
Biloculina depressa d'Orb. var. serrata, B r a d y, 1884, pl. III, fig. 3.
Pyrgoella sphaera (d'O r b i g n y)
Biloculina sphaera d'Orbigny, B r a d y, 1884, pl. II, fig. 4.
Pyrulina cylindroides (R o e m e r)
Polymorphina lanceolata Reuss, B r a d y, 1884, pl. LXXII, figs. 5, 6.
Robertina bradyi C u s h m a n & P a r k e r
Bulimina subteres Brady, B r a d y, 1884, pl. L, fig. 18.
Robertina bradyi Cushman & Parker, P h l e g e r, P a r k e r & P e i r s o n, 1953, pl. 6, figs. 19, 20.
Rotamorphina laevigata (P h l e g e r & P a r k e r)
Rotamorphina laevigata (Phleger & Parker), P h l e g e r, P a r k e r & P e i r s o n, 1953, pl. 10, figs. 17, 23.
Saracenaria italica D e f r a n c e
Cristellaria italica (Defrance), B r a d y, 1884, pl. LXVIII, figs. 17, 18, 20-23.
Siphouvigerina auberiana (d'O r b i g n y)
Uvigerina asperula Czjzek, B r a d y, 1884, pl. LXXV, figs. 6-8;
Uvigerina asperula var. auberiana d'Orbigny, B r a d y, 1884, pl. LXXV, fig. 9;
Uvigerina asperula var. ampullacea B r a d y, 1884, pl. LXXV, figs. 10, 11;
Uvigerina auberiana d'Orbigny and variant, P h l e g e r, P a r k e r & P e i r s o n, 1953, pl. 7, figs. 30-35.
Sphaeroidina bulloides d'O r b i g n y
Sphaeroidina bulloides d'Orbigny, B r a d y, 1884, pl. LXXXIV, figs. 1-7.
Spirillina decorata B r a d y
Spirillina decorata B r a d y, 1884, pl. LXXXV, figs. 22-25.
Spirosigmoilina tenuis (C z j z e k)
Spiroloculina tenuis (Czjzek), B r a d y, 1884, pl. X, figs. 7-9, 11;
Sigmoilina tenuis (Czjzek), P h l e g e r, P a r k e r & P e i r s o n), 1953, pl. 5, fig. 18.
Trifarina angulosa (W i l l i a m s o n)
Uvigerina angulosa Williamson, B r a d y, 1884, pl. LXXIV, figs. 15, 16.
Triloculina tricarinata d'O r b i g n y
Miliolina tricarinata d'Orbigny, B r a d y, 1884, pl. III, fig. 17.
Vaginulina spinigera B r a d y
Vaginulina spinigera Brady, B r a d y, 1884, pl. LXVII, figs. 13, 14.
Valvulineria arctica G r e e n
Valvulineria arctica G r e e n, 1960, pl. 1, fig. 3.
Valvulineria complanata (d'O r b i g n y)
Valvulineria complanata (d'Orbigny), P a r k e r, 1958, pl. 3, figs. 42-44.

iv. *Displaced benthonic species*

Ammonia beccarii (L i n n é)
Rotalia beccarii (Linné), B r a d y, 1884, pl. CVII, figs. 2, 3.
"Rotalia" beccarii (Linné) variants, P h l e g e r, P a r k e r & P e i r s o n, 1953, pl. 9, figs. 12-15, 17, 18.
Amphistegina lessonii D'O r b i g n y
Amphistegina lessonii d'Orbigny, B r a d y, 1884, pl. XCI, figs. 1-7;
P h l e g e r, P a r k e r & P e i r s o n, 1953, pl. 9, figs. 28, 29.
Asterigerinata mamilla (W i l l l i a m s o n)
Rotalina mamilla W i l l i a m s o n, 1858, pl. IV, figs. 109-111.
Cancris oblongus (W i l l i a m s o n)
Pulvinulina oblonga (Williamson), B r a d y, 1884, pl. CVI, fig. 4;
Cancris. cf. oblonga (Williamson), P h l e g e r, P a r k e r & P e i r s o n, 1953, pl. 9, figs. 26, 27.
Cibicides ex. gr. *lobatulus* (W a l k e r & J a c o b)
For references to figures see sub (iii). Only suspected displaced specimens are recorded here.
Cribroelphidium gunteri (C o l e)
Elphidium gunteri C o l e, 1931, pl. IV, figs. 9-10.
Cribrononion advenum (C u s h m a n)
Polystomella subnodosa (Münster), B r a d y, 1884, pl. XC, fig. 1;
Elphidium advenum (Cushman), P h l e g e r, P a r k e r & P e i r s o n, 1953, pl. 6, fig. 15.
Cribononion incertum (W i l l i a m s o n)
Polystomella striatopunctata (Fichtel & Moll), B r a d y, 1884, pl.CIX, figs. 22, 23.
Elphidium crispum (L i n n é)
Polystomella crispa (Linné), B r a d y, 1884, pl. CX, figs. 6, 7.
Elphidium crispum (Linné), P h l e g e r, P a r k e r & P e i r s o n, 1953, pl. 6, fig. 17.
Eponides repandus (F i c h t e l & M o l l)
Pulvinulina repanda (Fichtel & Moll), B r a d y, 1884, pl. CIV, fig. 18.
Florilus asterizans (F i c h t e l & M o l l)

Nonionina boueana d'Orbigny, B r a d y, 1884, pl. CIX, figs. 12, 13.
Nonion asterizans (Fichtel & Moll), P h l e g e r, P a r k e r & P e i r s o n, 1953, pl. 6, fig. 3.

Florilus atlanticus (C u s h m a n)
Nonionella atlantica Cushman, P h l e g e r, P a r k e r & P e i r s o n, 1953, pl. 6, figs. 9, 10.

Hanzawaia ex. gr. *concentrica* (C u s h m a n)
Cibicides concentrica (Cushman), C u s h m a n, 1930, pl. XII, fig. 4.

Neoconorbina orbicularis (T e r q u e m)
Discorbina orbicularis (Terquem), B r a d y, 1884, pl. LXXXVIII, figs. 4-8.

Pararotalia armata (d'O r b i g n y)
Calcarina armata (d'Orbigny), P a r k e r, J o n e s & B r a d y, 1865, pl. III, fig. 88.

Planorbulina mediterranensis d'O r b i g n y
Planorbulina mediterranensis d'Orbigny, B r a d y, 1884, pl. XCII, figs. 1-3; P h l e g e r, P a r k e r & P e i r s o n, 1953, pl. 11, figs. 20, 21.

Poritextularia panamensis (C u s h m a n)
Textularia panamensis C u s h m a n, 1918, pl. CC, fig. 1.

Ptychomiliola separans (B r a d y)
Miliolina separans Brady, B r a d y, 1884, pl. VII, figs. 1-4.

Reussella miocenica C u s h m a n
Verneuilina spinulosa Reuss, B r a d y, 1884, pl. XLVII, figs. 1-3.

Rosalina globularis d'O r b i g n y
Discorbina globularis (d'Orbigny), B r a d y, 1884, pl. LXXXVI, fig. 13.

REFERENCES

Alexandrowicz, S.W. (1963) – Stratigraphy of the Miocene deposits in the Upper Silesian Basin. Prace Instytutu Geol., tom. 39.

Anikouchine, W.A. & Hsin-Yi-Ling (1967) – Evidence for turbiditic accumulation in trenches in the Indo-Pacific region. Marine Geology, vol. 5.

Arrhenius, G. (1963) – Pelagic sediments. In: The Sea, M.N. Hill (editor), Interscience, vol. 3.

Ascoli, P. (1958) – Studio micropaleontologico preliminare sulla posizione stratigrafica della cosidetta "Pietra di Cantoni" nel Tortonese. Boll. Soc. Geol. It., vol. LXXVII.

Bandy, O.L. (1961) – Distribution of Foraminifera, Radiolaria and diatoms in sediments of the Gulf of California. Micropaleontology, vol. 7, no. 1.

——— (1964) – Foraminiferal trends associated with deepwater sands, San Pedro and Santa Monica basins, California. J.P., vol. 38, no. 1.

——— (1967) – Foraminiferal definition of the boundaries of the Pleistocene in Southern California, U.S.A. Progress in Oceanography, vol. 4.

Bandy, O.L. & M.A. Chierici (1966) – Depth-temperature evaluation of selected California and Mediterranean bathyal Foraminifera. Marine Geology, vol. 4.

Bandy, O.L., E. Frerichs & E. Vincent (1967) – Origin, development, and geological significance of Neogloboquadrina Bandy, Frerichs and Vincent, gen. nov. C.C.F.F.R., vol. 19, pt. 4.

Banner, F.T. & W.H. Blow (1967) – The origin, evolution and taxonomy of the foraminiferal genus Pulleniatina, Cushman 1927. Micropaleontology, vol. 13, no. 2.

Bé, A.W.H. & W.H. Hamlin (1967) – Ecology of Recent planktonic Foraminifera. Part 3 – Distribution in the North Atlantic during the summer of 1963. Micropaleontology, vol. 13, no. 1.

Berger, W.H. (1967) – Foraminiferal ooze: Solution at Depths. Science, vol. 156.

——— (1969) – Ecologic patterns of living planktonic Foraminifera. Deep-Sea Research, vol. 16.

——— (1970) – Planktonic Foraminifera: Selective solution and the Lysocline. Marine Geology, vol. 8.

Berggren, W.A. (1968) – Micropaleontology and the Pliocene/Pleistocene boundary in a deep-sea core from the South Central North Atlantic. Giornale di Geologia (2), vol. XXXV, fasc. II.

Berggren, W.A., J.D. Phillips, A. Bertels & D. Wall (1967) – Late Pliocene – Pleistocene stratigraphy in deep sea cores from the south-central North Atlantic. Nature, vol. 216.

Berthois, L., A. Crosnier & Y. Le Calvez (1968) – Contribution à l'étude sedimentologique du Plateau Continental dans la Baie de Biafra. Cah. O.R.S.T.O.M., Sér. Océanogr., vol. VI, no 3-4.

Blow, W.H. (1959) – Age, correlation and biostratigraphy of the upper Tocuyo (San Lorenzo) and Pozón Formations, eastern Falcon, Venezuela. Bull. Amer. Pal., vol. 39, no. 208.

Bolli, H.M. (1957) – Planktonic Foraminifera from the Oliogene-Miocene Cipero and Lengua Formations of Trinidad, B.W.I. U.S.N.M., Bull. 215.

——— (1964) – Observations on the stratigraphic distribution of some warm water planktonic Foraminifera in the Young Miocene to Recent. Ecl. Geol. Helv., vol. 57, no. 2.

Bolli, H.M. & P.J. Bermudez (1965) – Zonation based on planktonic Foraminifera of Middle Miocene to Pliocene warm-water sediments. Bol. Inf. Asoc. Ven. Geol., Min. Petr., vol. 8, no. 5.

Brady, H.B. (1884) – Report on the Foraminifera dredged by HMS Challenger, during the years 1873-1876. Report Scientific Results Explor. Voyage HMS Challenger. Zoology, vol. 9.

Brouwer, J. (1965) – Agglutinated foraminiferal Faunas from some turbiditic sequences I-II. K.N.A.W., Proc., ser. B, vol. 68, no. 5.

——— (1967) – Foraminiferal faunas from a graded-bed sequence in the Adriatic Sea. K.N.A.W., Proc., ser. B, vol. 70, no. 3.

Bruun, A.F. (1953) – Animal life of the deep sea bottom. The Galathea Deep Sea Expedition, 1950-1952. George Allen and Unwin Ltd. London.

Calvert, S.E. (1968) – Silica balance in the ocean and diagenesis. Nature, vol. 219.

Cifelli, R. & K.N. Sachs (1966) – Abundance relationship of planktonic Foraminifera and Radiolaria. Deep-Sea Research, vol. 13.

Cita, M.B. & S. d'Onofrio (1967) – Climatic fluctuations in submarine cores from the Adriatic Sea (Mediterranean). Progress in Oceanography, vol. 4.

Cole, W.S. (1931) – The Pliocene and Pleistocene Foraminifera of Florida. Florida State Geol. Survey, Bull. 6.

Correns, C.W. (1937) – Die Sedimente des äquatorialen Atlantischen Ozeans. Wiss. Ergebn. Deut. Atl. Exped. "Meteor", 1925-1927, vol. 3, pt. 3.

Cushman, J.A. (1918) – The smaller fossil Foraminifera of the Panama Canal Zone. USNM, Bull. 103.

——— (1922) The Foraminifera of the Atlantic Ocean. Pt. 3. USNM, Bull. 104, pt. 3.

——— (1930) – The Foraminifera of the Choctawhatchee formation of Florida. Florida State Geol. Survey, Bull. 4.

——— (1931) – The Foraminifera of the Atlantic Ocean, Pt. 8. USNM, Bull. 104, pt. 8.

Davies, D.K. (1968) – Carbonate turbidites, Gulf of Mexico. J. Sed. Petrol., vol. 38, no. 4.

Dietz, R.S. (1964) – Origin of continental slopes. American Scientist, vol. 52.

d'Orbigny, A.D. (1846) – Foraminifères fossiles du Bassin Tertiaire de Vienne (Autriche). Gide et Comp., Paris.

Earland, E.A. (1936) – Foraminifera. Part IV. Additional records from the Weddell Sea sector from material obtained by the S.Y. "Scotia". Discovery Reports, vol. 13.

Emery, K.O., J. Hülsemann & K.S. Rodolfo (1962) – Influence of turbidity currents upon basin waters. Limn. and Oceanog., vol. 7, no. 4.

Ericson, D.B. (1963) – Cross-correlation of deep-sea sediment cores and determination of relative rates of sedimentation by micropaleontological techniques. In: The Sea, M.N. Hill (editor), Interscience, vol. 3.

Ericson, D.B., M. Ewing & G. Wollin (1963) – Pliocene-Pleistocene boundary in deep-sea sediments. Science, vol. 39, no. 3556.

——— (1963) The Pleistocene epoch in deep-sea sediments. Science, vol. 146, no. 3645.

Ericson, D.B., M. Ewing, G. Wollin & B.C. Heezen (1961) – Atlantic deep-sea sediment cores. Bull. Geol. Soc. Amer., vol. 72.

Ericson, D.B. & G. Wollin (1956) – Correlation of six cores from the equatorial Atlantic and the Caribbean. Deep-Sea Research, vol. 3.

Eckert, H.R. (1965) Une station d'observation sur les foraminifères planktoniques actuels dans le Golfe de Guinée. Ecl. Geol. Helv., vol. 58, no. 2.

Frerichs, W.E. (1967) – Distribution and ecology of Foraminifera in the sediments of the Andaman Sea. Dissertation: Univ. S.Calif.

——— (1971) – Paleobathymetric trends of Neogene foraminiferal assemblages and sea floor tectonism in the Andaman Sea area. Marine Geology, vol. 11, no. 3.

Funnell, B.M. (1967) – Foraminifera and Radiolaria as depth indicators in the marine environment. Marine Geology, vol. 5, nos. 5/6.

Green, K.E. (1960) – Ecology of some Arctic foraminifera. Micropaleontology, vol. 6, no. 1.

Griffin, J.J., H. Windom & E.D. Goldberg (1968) – The distribution of clay minerals in the world ocean. Deep-Sea Research, vol. 15.

Griggs, G.B. & G.A. Fowler (1971) – Foraminiferal trends in a Holocene turbidite. Deep-Sea Research, vol. 18, no. 6.

Grimsdale, T.F. & F.P.C.M. van Morkhoven (1955) – The ratio between pelagic and benthonic Foraminifera as a means of

estimating depth of deposition of sedimentary rocks. Proc. 4th W.P.C., section I/D, Reprint 4.
Grunau, H.R. (1965) – Radiolarian cherts and associated rocks in space and time. Ecl. Geol. Helv., vol. 58, no. 1.
Heezen, B.C. (1963) – Turbidity currents. In: The Sea, M.N. Hill (editor), Interscience, vol. 3.
Houbolt, J.J.H.C. (1973) – The deep-sea canyons in the Gulf of Guinea near Fernando Poo. (in this volume)
Huang, T. (1964) – Smaller Foraminifera from the Sanshien-Chi, Taitung, Eastern Taiwan. Proc. Geol. Soc. China, no. 7.
Hudson, J.D. (1967) – Speculations on the depth relation of calcium carbonate solution in recent and ancient seas. Marine Geology, vol. 5, nos. 5/6.
Il'In, A.V., I.I. Shurko, D.S. Nikolayev & Ye.I. Yefimova (1967) – Stratification of abyssal sediments in the equatorial Atlantic. Dokl. Akad. Nauk SSSR, Earth Science Sections, 176.
Keyzer, F.G. (1953) – Reconsideration of the so-called Oligocene fauna in the asphaltic deposits of Buton (Malay Archipelago).
Kornicker, L.S. (1959) – Analysis of factors affecting quantitative estimates of organism abundance. J. Sed. Petrol., vol. 29, no. 4.
Lagaaij, R. (1972) – Shallow-water Bryozoa from deep-sea sands of the Principe Channel, Gulf of Guinea. Proc. 2nd Intern. Conf. on Bryozoa, I.B.A., Durham, 1971 (in press).
Lagaaij, R. (1973) – Shallow-water Bryozoa from deep-sea sands of the Principe Channel, Gulf of Guinea. In: Living and Fossil Bryozoa, Recent Advance, in research, G.P. Larwood Ed. Academic Press London 1973.
région d'Abidjan (Côte d'Ivoire). Rev. Micropal., vol. 6, no. 1.
Leroy, L.W. (1939) – Some small Foraminifera, Ostracoda and Otoliths from the Neogene ("Miocene") of the Rokan-Tapanoeli Area, Central Sumatra. Natuurk. Tijdschr. Nd. Indië, vol. 99, no. 6, pt. 3.
——(1941) – a, Small Foraminifera from the Late Tertiary of the Sangkulirang Bay area, East Borneo. Colorado School of Mines Quarterly, vol. 36, no. 1, part 1.
——(1941) – b, Small Foraminifera from the Late Tertiary of Siberoet Island, off the West Coast of Sumatra. Colorado School of Mines, Quarterly, vol. 36, no. 1, part 2.
Ling, H.Y. & W.A. Anikouchine (1967) – Some Spumellarian Radiolaria from the Java, Philippine and Mariana Trenches. J.P., vol. 41, no. 6.
Lipps, J.H. (1965) – Revision of the foraminiferal family Pseudoparrellidae Voloshinova. Tulane studies in Geol., vol, 3, no. 2..
Loeblich, A.R. & H. Tappan (1964) – Sarcodina, chiefly "Thecamoebians" and Foraminiferida. Treatise on Invertebrate Paleontology, part C, Prostista 2, vols. 1 and 2.
Murray, J. & Renard (1891) – Report on deep-sea deposits based on the specimens collected during the voyage of HMS Challenger in the years 1872 to 1876. Report Scientific Results Explor. voyage HMS Challenger.
Ogniben, L. (1958) – Stratigrafia e microfauna del Terziario della zona di Caiazzo (Caserta). Riv. It. Pal. Strat., vol. LXIV, nos. 2-3.
Ovey, C.D. (1952) – On the validity and use of planktonic Foraminifera in the interpretation of past climatic changes from a study of deep-sea cores. Geol. Rundschau, vol. 40, no. 1.
Parker, F.L. (1954) – Distribution of the Foraminifera in the northeastern Gulf of Mexico. Bull. Mus. Comp. Zool., vol. 111, no. 10.
——(1958) – Eastern Mediterranean Foraminifera. Reports Swedish Deep-Sea Exped., vol. 8, pt. 4.
Parker, F.L. & W.H. Berger (1971) – Faunae and solution patterns of planktonic Foraminifera in surface sediments of the South Pacific. Deep-Sea Research, vol. 18.
Parker, W.K., T.R. Jones & H.B. Brady (1865) – On the nomenclature of the Foraminifera. Pt. 12. The species enumerated by d'Orbigny in the "Annales des sciences Naturelles", vol. 7, 1826. Ann. & Mag. Nat. History, ser. 3, vol. 16.
Petrovic, M.V. (1962) – Beitrag zur Kenntis der Mikrofauna aus der Umgebung von Stubik, Veliki Izvor und Vojilovo. Bull. Mus. Hist. Nat. Belgrade, ser. A, livr. 14-15.
——(1962) – Beitrag zur Kenntis der Tortonischen Foraminiferen von Beograd und ihrer näheren Umgebung. Ann. Geol. de la Penin. Balkan, Inst. Geol. Univ. Beograd, t. 29.
Petters, V. & R. Sarmiento (1956) – Oligocene and Lower Miocene biostratigraphy of the Carmen-Zambrano area, Colombia. Micropaleontology, vol. 2, no. 1.
Phleger, F.B. (1951) – Displaced Foraminifera faunas. SEPM, Spec. Publ. 2.
——(1955) – Foraminiferal faunas in cores offshore from the Mississippi Delta. Deep-Sea Research, Suppl. to vol. 3.
Phleger, F.B. & F.L. Parker (1951) – Ecology of foraminifera. Northwest Gulf of Mexico. Geol. Soc. Amer., Memoir 46.
Phleger, F.B., F.L. Parker & J.F. Peirson (1953) – North Atlantic Foraminifera. Reports Swedish Deep-Sea Exped., vol. 7, no. 1.
Piggot, C.S. (1944) – Radium content of ocean-bottom sediments. Carnegie Inst. of Wash., Publ. 556.
Posthuma, J.A. (1971) – Manual of planktonic Foraminifera. Elsevier Publishing Company.
Pytkowics, R.M. (1965) – Calcium carbonate saturation in the Ocean. Limnol. Oceanogr., vol. 10.
Ramsay, A.T.S. (1971) – Occurrence of Biogenic Siliceous Sediments in the Atlantic Ocean. Nature, vol. 233, no. 5315.
Revelle, R.R. (1944) – Marine bottom samples collected in the Pacific Ocean by the Carnegie on its seventh cruise. Carnegie Inst. of Wash., Publ. 556.
Riedel, W.R. (1951) – Number of Radiolaria in sediments. Nature, vol. 167.
——(1963) – The preserved record: Paleontology of pelagic sediments. In: The Sea, M.N. Hill (editor), Interscience, vol. 3.
Ruddiman, W.F. & B.C. Heezen (1967) – Differential solution of Planktonic Foraminifera. Deep-Sea Research, vol. 14.
Saidova, H.M. (1961) – Ekologija i Paleogeografija Dal'nevostocknykh morej SSSR i Severo-Zapadnoj Chasti Tikhogo Okeana. Akad. Nauk SSSR, Inst. Okean., 1961, pp. 1-232.
——(1964) – Distribution of bottom Foraminifera and stratigraphy of sediment in northeastern Pacific. Trudy Instyt. Okean. Akad, Nauk SSSR.
Schott, W. (1937) – Die Foraminiferen in dem äquatorialen Teil des Atlantischen Ozeans. Wiss. Ergebn. Deut. Atlant. Exped. "Meteor", 1925-1927, vol. 3, pt. 3.
——(1952) – On the sequence of deposits in the equatorial Atlantic Ocean. Handl. Göteborgs Kungl. Vetensk. och Vitterhets Samh., följ. 7, ser. B, vol. 6, no. 2.
Stschedrina, Z.G. (1947) – On the distribution of the Foraminifera in the Greenland Sea. Dokl. Akad. Nauk. SSSR, vol. 55, no. 9.
Sverdrup, H.U., M.W. Johnson & R.H. Fleming (1946) – The Oceans. Prentice Hall, New Jersey.
Tavares Rocha, A. & L.M. Ubaldo (1964) – Foraminiferos do Terciario superior e do Quaternario da Provincia Portuguesa de Timor. Mem. Junta Invest. Ultramar, no. 51.
Tesch, J.J. (1948) – The thecosomatous pteropods: The Atlantic. Dana Reports, vol. 28.
Todd, R. (1958) – Foraminifera from western Mediterranean deep-sea cores. Reports Swedish Deep-Sea Exped., vol. 8, pt. 3.
Walton, W.R. (1964) – Recent foraminiferal ecology and paleoecology. In: J. Imbrie & N. Newell (editors): Approaches to Paleoecology, Wiley, N. York.
Williamson, W.C. (1858) – On the Recent Foraminifera of Great Britain. Ray Soc. Publs.

MIX
Papier aus verantwortungsvollen Quellen
Paper from responsible sources
FSC® C105338

If you have any concerns about our products,
you can contact us on
ProductSafety@springernature.com

In case Publisher is established outside the EU,
the EU authorized representative is:
Springer Nature Customer Service Center GmbH
Europaplatz 3, 69115 Heidelberg, Germany

Printed by Libri Plureos GmbH
in Hamburg, Germany